Land Use in Product Life Cycles and Ecosystem Quality

European University Studies
Europäische Hochschulschriften
Publications Universitaires Européennes

Series V
Economics and Management

Reihe V Série V
Volks- und Betriebswirtschaft
Sciences économiques

Vol./Band 3012

PETER LANG
Bern · Berlin · Bruxelles · Frankfurt a. M. · New York · Oxford · Wien

Thomas Köllner

Land Use in Product Life Cycles and Ecosystem Quality

PETER LANG
Bern · Berlin · Bruxelles · Frankfurt a. M. · New York · Oxford · Wien

Bibliographic information published by Die Deutsche Bibliothek
Die Deutsche Bibliothek lists this publication in the Deutsche Nationalbibliografie; detailed bibliographic data is available on the Internet at ‹http://dnb.ddb.de›.

British Library and Library of Congress Cataloguing-in-Publication Data:
A catalogue record for this book is available from *The British Library*, Great Britain, and from *The Library of Congress*, USA

This work was accepted as a doctoral thesis by the University of St. Gallen, Institute for Economy and the Environment.

ISSN 0531-7339
ISBN 3-03910-207-9
US-ISBN 0-8204-6870-3

Preface

A prerequisite of environmental decision-making is the ability to assess the environmental consequences of alternative decisions. The basic knowledge that is drawn upon in such an assessment is generated in the environmental sciences. However, the context in which such decisions are made is one where access to such basic knowledge is limited. Life Cycle Assessment (LCA), which is a decision support tool, can potentially bridge that gap between environmental sciences and decision-makers.

Being a biologist by training with interdisciplinary interests, my objective was to sample the knowledge and the information on ecological impacts of land use and then condense these into a method for the assessment of land use within LCA. The added value for decision-makers is that access to environmental knowledge is provided in such a manner that they can easily consider it in the decision process. However, the information loss, which results from the condensing process, is large. From the ecologists' point of view the method proposed in this book and the generated results might appear to be coarse and preliminary. However, from the decision-makers' point of view the results might be very valuable for improving their decisions. To conduct LCAs in industry, government, and other organisations, commercial software is used. This means that integrating my results into such software allows potential users to indirectly access my research. Fortunately, I was able to actively participate in a Dutch/Swiss project team in order to help develop the land use part of the Eco-indicator 99 (www.pre.nl). This is a method of Life Cycle Impact Assessment, which is, in turn, part of commercial LCA software packages like SimaPro.

This book is a result of my research, carried out between 1996 and 2000 at the Institute for Economy and the Environment, University St. Gallen (IWÖ) and at the Natural and Social Science Interface, Swiss Federal Institute of Technology Zurich (UNS).

For their confidence in my research capability and for coaching me in my Ph.D. project, I would like to thank Ruedi Müller-Wenk (IWÖ) and Thomas Dyllick (IWÖ). It was Thomas Dyllick who stimulated my effort to keep the potential users of my assessment method in mind and not forget the users' needs when diving into the numerical modelling. At the same time, I appreciated Ruedi Müller-Wenk's valuable feedback concerning the detailed development of the method for assessing land use. I am also grateful to Roland Scholz (UNS) and Olaf Tietje (UNS) for providing me the opportunity to work in the very motivating environment of the UNS as well as for their continuous help whenever needed.

Many other people also supported my research through motivating discussions, offering critical comments, and providing data. For all this I want to explicitly thank Thomas Baumgartner (UNS), Arthur Braunschweig (IWÖ), Thomas Dalang (WSL), Mark Goedkoop (Pré Consultants), Patrick Hofstetter (U.S. EPA), Niels Jungbluth (ESU-Services), Erwin Lindeijer (TNO Industrie), Thomas Mettier (UNS), Klaus Richter (EMPA) and Thomas Wohlgemuth (WSL). Thomas Wohlgemuth and Thomas Dalang in particular provided much of the empirical data on species diversity and vegetation types in Switzerland. Niels Jungbluth provided the inventory data for the case study.

In addition, I was able to participate in two working groups addressing the issue of land use in LCA. One group, named *Abiotic Resources and Land Use,* is a part of SETAC's working group on impact assessment. The other is the working group *Land Use* which is a subgroup of the COST Action E9 *Life Cycle Assessment for Forestry and Forest Products*. Thanks also to Laura Cohen who improved my written English considerably. I would also like to express my thanks for the financial support provided me by the Swiss National Science

Foundation within the Priority Programme Environment (Project: *Environmental Prioritizing* as part of *Society I – Sustainable Nutrition: Visions, Strategies, and Instruments*) and by the Swiss Office for Education and Science (Project: *Land use in Life Cycle Assessment*).

Thomas Köllner
Zürich in August 2003

Table of contents

Summary

Land use is an economic activity that generates large benefits for human society. However, as a side effect it has caused many environmental problems throughout history and still does today. Some examples of these are endangerment of species, nutrification of lakes and rivers, loss of fertile soil, and impact on landscapes' aesthetics. All of these impacts put pressure on regional ecosystem quality. In this book, I focused on biodiversity, which is conceptualised as an important part of ecosystem quality. International treaties – the most important being the Convention on Biodiversity – stress biodiversity's intrinsic value as well as its value for the functioning of the world's ecosystem.

Life Cycle Assessment (LCA) is a tool for addressing such environmental problems. It is particularly used for environmental planning and decision-making with respect to the whole life cycle of a product. However, land use and its consequences for ecosystem quality still aren't adequately taken into account in LCA. My objectives were (i) to develop a method of assessing land use impacts on biodiversity which is suitable for LCA, (ii) to make the general method operable using empirical information on species diversity, and (iii) to apply calculated characterisation factors in a simple case study.

In Life Cycle Impact Assessment (LCIA), characterisation factors are used to weight an environmental intervention with respect to its potential to damage the defined safeguard subject. To develop characterisation factors for land use, a modelling approach was applied. The modelling was split up into different stages. First, a qualitative model was developed linking land use activities and impacts on ecosystem quality. The rather general safeguard subject *ecosystem quality* was split up into more specific attributes (biodiversity, ecosystem functions, and ecological resources). Next, land use impacts on local and regional species diversity were quantified. The use of effect-damage curves allowed observed effects on species diversity to be transformed into potential damage/benefits. The characterisation factor *ecosystem damage potential* ($EDP^{sp\text{-}div}$) refers to those damage/benefits. Finally, local and regional damage potentials were condensed into a single factor.

I developed a set of diverse characterisation factors $EDP^{sp\text{-}div}$, which made it possible to take into account different background situations on the regional scale. In addition differences in sensitivity to land use activities of all of the species and of those species which are already threatened could be considered. The characterisation factors were calculated according to a linear as well as a nonlinear model. Data uncertainties were assessed with a Monte-Carlo simulation.

The results of the analysis show that land use types differ remarkably in their local species richness and corresponding local $EDP^{sp\text{-}div}$ value. Detrimental land use types with low species numbers (0-20 species/0.01 ha) are continuous urban area, conventional arable land, intensive meadow, and sports facilities. Moderate types (20-40 species/0.01ha) are semi-natural broad-leafed forest, organic arable land, and green urban area. Beneficial types (40-60 species/0.01ha) are industrial fallow, agricultural fallow, organic meadow and forest edges. Regional $EDP^{sp\text{-}div}$ factors show a similar picture (high- and low-intensity artificial land are very detrimental to regional species richness, high-intensity agriculture and high-intensity forestry are detrimental, low-intensity agriculture and low-intensity forestry are beneficial). Overall, characterisation factors were calculated for 30 land use types and eight types of intensity of land use, allowing a varied impact assessment of land use.

The application of developed $EDP^{sp\text{-}div}$ factors is rather simple. In order to calculate the total damage of a land use activity, the area of the specific land use type, its duration of use, and its specific characterisation factor are multiplied. An exemplary case study assessing the land use of agricultural products was conducted for vegetables. Vegetables' life cycle consists of 4 phases: production, storage, transportation, and packaging. Application of the characterisation factors resulted in a change of the ranking of the alternatives investigated.

For operational calculations of $EDP^{sp\text{-}div}$, plant species diversity was chosen as the indicator for biodiversity. Although taking other species groups into account in addition could change the detailed results, the overall ranking of the land use types is expected to be robust. If threatened species were weighted more than non-threatened species, then near-to-natural land use types (e.g., natural grassland) would benefit in the ranking. In general, the local $EDP^{sp\text{-}div}$ factors calculated can be applied to the assessment of land use in Western European countries. However, the validity of the results decreases as the distance to the study region (Swiss Plateau, Germany) increases. In contrast, the regional factor is based on a regression analysis using data from the Swiss Alps therefore, application of the results is limited and should only be used for alpine regions.

Keeping the limitations described above in mind, one can conclude that this book improves damage-oriented methods for Life Cycle Impact Assessment, such as the Eco-indicator 99, in assessing land use. However, in the future it would be advisable to include empirical information covering a wide geographical range.

Zusammenfassung

Landnutzung ist eine ökonomische Aktivität, die grossen Nutzen generiert. Gleichzeitig war Landnutzung in der Vergangenheit, und ist dies heute noch, Auslöser für viele Umweltprobleme. Beispiele hierfür sind das Artensterben, die Eutrophierung von Oberflächengewässern, der Verlust von fruchtbarem Ackerboden und die negativen Einflüsse auf die Ästhetik von Landschaften. Alle diese Einflüsse setzen das regionale Ökosystem unter Druck. In diesem Buch habe ich mich jedoch auf die Betrachtung der Biodiversität beschränkt, welche als wichtiger Bestandteil von Ökosystemqualität konzeptionalisiert wurde. Internationale Verträge – der wichtigste in diesem Zusammenhang ist zweifelsohne die *Konvention zum Schutz der Biodiversität* – betonen sowohl den intrinsischen Wert von Biodiversität, als auch ihr Wert für das Funktionieren der Ökosysteme weltweit.

Gleichzeitig stehen Instrumente, wie die Produktökobilanz, zur Verfügung, die Umweltprobleme thematisieren. Die Produktökobilanz wird insbesondere für die Planung und Entscheidungsfindung im Umweltbereich eingesetzt. Dabei wird auf den gesamten Lebenszyklus, d.h. einschiesslich der Schritte Rohstoffgewinnung, Produktion, Transport, Konsum und Entsorgung, eines Produktes und dessen Folgen für die Umwelt Bezug genommen. Landnutzung ist jedoch mit ihren Folgen für Ökosystemqualität bisher kaum in Ökobilanzen berücksichtigt. Meine Ziele waren daher (i) eine Methode zu entwickeln, die es erlaubt Einflüsse von Landnutzung auf Biodiversität in der Ökobilanz fassbar zu machen, (ii) die allgemeine Methode unter Rückgriff auf empirische Information zur Artenvielfalt zur Anwendungsreife zu bringen und (iii) diese in einer einfachen Fallstudie anzuwenden.

In der Ökobilanz werden als zentraler Bestandteil Charakterisierungsfaktoren benutzt, um verschiedene Umwelteinwirkungen (z.B. Emissionen) zu gewichten, und zwar nach Massgabe ihres Schädigungspotentials bezüglich vorher definierte Schutzgüter. Um diese Faktoren zu entwickeln wurde ein Modellierungsansatz gewählt. Die Modelentwicklung gliedert sich in verschiedene Abschnitte. Zuerst wurde ein qualitatives Model entworfen um Landnutzungsaktivitäten und deren Einflüsse auf Ökosystemqualität zu fassen. Das relativ allgemeine Schutzgut Ökosystemqualität wurde für die Operationalisierung in spezifische Attribute (Biodiversität, ökologische Funktionen und ökologische Ressourcen) aufgeteilt. Als nächstes wurden Auswirkungen der Landnutzung auf lokale und regionale Artenvielfalt modelliert. Wirkungs-Schadens-Kurven erlaubten die Auswirkungen auf Artenvielfalt in potentielle Schäden bzw. Nutzen zu transformieren. Diese entsprechen den Charakterisierungsfaktoren, die *ecosystem damage potential* ($EDP^{sp\text{-}div}$) genannt wurden. Zum Schluss wurden der lokale und der regionale Faktor in einen einzelnen Faktor zusammengefasst.

Es wurden verschiedene Charakterisierungsfaktoren entwickelt, um Unterschiede in der regionalen Belastungssituation und Unterschiede in der Sensitivität zwischen allen Arten insgesamt und den Arten, die bedroht sind, berücksichtigen zu können. Die Faktoren wurden mit einem linearen und einem nicht linearen Model berechnet. Datenunsicherheiten wurden mit einer Monte-Carlo Analyse modelliert.

Die Ergebnisse zeigen, dass Landnutzungstypen stark in ihrer lokalen Artenvielfalt und entsprechendem $EDP^{sp\text{-}div}$ variieren können. Ungünstige Landnutzungstypen mit niedrigen Artenzahlen (0-20 Arten/0.01 ha) sind kontinuierliche Siedlungsfläche, konventionell bewirtschaftete Ackerfläche, intensiv bewirtschaftete Wiese und Sportplätze. Mittelmässig (20-40 Arten/0.01 ha) sind halbnatürliche Laubwälder, organisch bewirtschaftete Ackerflächen und städtische Grünanlagen. Günstige (40-60 Arten/0.01 ha) sind dagegen Industriebrachen, Ackerbrachen, organisch bewirtschaftete Wiesen und Waldränder. Die regionalen $EDP^{sp\text{-}div}$ Faktoren zeigen ein ähnliches Bild: Siedlungsflächen sind sehr ungünstig, intensive Forst- und Landwirtschaft sind ungünstig, dagegen sind extensive Forst- und Landwirtschaft günstig. Insgesamt wurden für 30 verschiedene Landnutzungstypen und acht Intensitätstypen Charakterisierungsfaktoren berechnet. Dies erlaubt eine differenzierte Beurteilung von Landnutzung in der Ökobilanz.

Die Anwendung der Charakterisierungsfaktoren ist relativ einfach. Um den endgültigen Schaden einer Nutzung durch einen bestimmten Landnutzungstyp zu berechnen, wird das Produkt aus genutzter Fläche und Zeitdauer mit dem spezifischen Charakterisierungsfaktor multipliziert. In einer einfachen Fallstudie wurde dies für die Landnutzung von Landwirtschaftprodukten (Gemüse) durchgeführt. Der einfache Lebenszyklus des Produktes Gemüse besteht aus 4 Abschnitten: Produktion, Lagerung, Transport und Verpackung. Die Fallstudie zeigte, dass die Anwendung der Charakterisierungsfaktoren die Rangreihenfolge der Alternativen verändern kann.

Für alle operativen Berechnungen wurde die Artenvielfalt von Gefässpflanzen als Indikator für Biodiversität gewählt. Obwohl die Berücksichtigung von anderen Artengruppen die Ergebnisse im Detail verändern kann, wird erwartet, dass die grobe Rangreihenfolge der Landnutzungstypen robust ist. Wenn jedoch bedrohten Arten ein stärkeres Gewicht zugesprochen werde würde, als den nicht bedrohten, dann müsste erwartet werden, dass naturnahe Landutzungstypen (naturnahe Wälder oder Wiesen) sich in der Rangreihenfolge verbessern würden. Im allgemeinen können die lokalen Charakterisierungsfaktoren für Westeuropa eingesetzt werden, jedoch mit zunehmender Entfernung zum Untersuchungsraum (Schweizer Mittelland, Deutschland) mit einer abnehmenden Validität. Im Gegensatz dazu sind die regionalen Faktoren zu einem gewissen Grad exemplarisch, da sie auf einer Regressionsanalyse mit Daten aus den Schweizer Alpen aufbauen und sollten insbesondere für alpine Regionen eingesetzt werden.

Unter Berücksichtigung aller Einschränkungen kann geschlossen werden, dass die vorliegende Arbeit schadensorientierte Bewertungsmethoden der Wirkungsbilanz wie den Eco-indicator 99 mit dem Element der Landnutzung ergänzen kann. Jedoch, ist es für die zukünftige Entwicklung wünschenswert, dass empirische Informationen von einem grösseren Einzugsgebiet berücksichtigt werden können.

Introduction

Life Cycle Assessment (LCA) is a method for evaluating the environmental burden of products or processes. The life cycle of a product is composed of many activities, such as resource exploitation, manufacturing, transportation, product use, and waste treatment (Fig. 1-1). The objective of LCA is to compare product alternatives in the context of the whole life cycle. Alternatives, which serve the same purpose (e.g., different types of packaging for beverages), are ranked with respect to their environmental burden. The results of LCA in combination with other information (e.g., costs of alternatives), support decisions in companies, governmental organisations, or non-governmental organisations.

Different input flows from the environmental system and output flows into the environmental system are assessed without regarding the time and space of their occurrence in the life cycle. The inputs are the resources needed during the different stages of the life cycle, such as ore, fossil oil, biomass, energy, and land. The outputs from the product system are the chemical emissions (e.g., toxic substances, nitrogen, SO_x, CO_2), noise, radiation, and solid waste. Physical changes of the landscape and soil are also possible outputs of the life cycle resulting from land use. In consequence these input and output flows cause many environmental impacts. These impacts are grouped into different impact categories, such as resource depletion, nutrification, acidification, human toxicity, and land use.

Land use is one impact category that is covered in Life Cycle Impact Assessment (LCIA) in a preliminary manner. Despite its global relevance (e.g., contribution to biodiversity decline), the environmental impact that results from land use is frequently omitted in Life Cycle Assessment. Characterisation factors suitable for ecological weighting of all relevant land use activities (e.g., agriculture, forestry, urban land) are missing. However, several proposals aim at characterising specific land use types (Frischknecht et al. 1996; Giegrich and Sturm 1996; Anonymus 1998). Only recently have researchers proposed methods for the characterisation that are applicable to all possible land use types simultaneously (e.g., Blonk et al. 1997; Lindeijer et al. 1998; Müller-Wenk 1998).

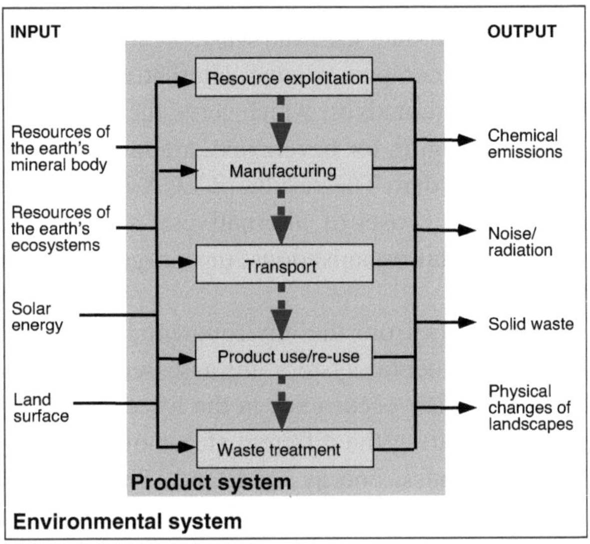

Fig: A product's life cycle with its flows of inputs and outputs.

Land use is one impact category that is covered in Life Cycle Impact Assessment (LCIA) in a preliminary manner. Despite its global relevance (e.g., contribution to biodiversity decline), the environmental impact that results from land use is frequently omitted in Life Cycle Assessment. Characterisation factors suitable for ecological weighting of all relevant land use activities (e.g., agriculture, forestry, urban land) are missing. However, several proposals aim at characterising specific land use types (Frischknecht et al. 1996; Giegrich and Sturm

1996; Anonymus 1998). Only recently have researchers proposed methods for the characterisation that are applicable to all possible land use types simultaneously (e.g., Blonk et al. 1997; Lindeijer et al. 1998; Müller-Wenk 1998).

The most straightforward approach to assess the environmental impacts that result from land use in LCIA is to simply add up the areas of concern and to multiply the sum of these areas by the duration of usage. However, it is obvious that one hectare of an intensively managed crop field does not result in the same environmental impact as one hectare of a natural forest aimed at wood production. Therefore, different land use types need to be weighted according to their different environmental impacts. The process of weighting implies that fundamental objectives are specified. The underlying assumptions are the existence of a utility measured on a cardinal scale and the compensability of subordinate utilities (see also Martinez-Alier et al. 1999). The different land use types are then evaluated according to the chosen objectives.

The fundamental objective, which is defined in LCIA, is to reduce damage to safeguard subjects (also called major categories of impacts or areas of protection) caused by anthropogenic impacts. In its most general form, these safeguard subjects are usually defined to be *ecosystems*, *human health,* and *resources* (e.g., Consoli et al. 1993; Lindfors et al. 1995; Udo de Haes et al. 1999). On a subordinate level, more specific attributes of the rather general safeguard subjects may be specified. Reducing damage to the attributes is then a subordinate objective.

My goal was to model the damage to ecosystem quality due to land use within the framework of LCA. It is quite important to acknowledge the framework of the modelling effort, because the type of decision to be supported by LCA determines the requirements (e.g., the degree of case specificity and site specificity).

More specifically, I consider an aspect of ecosystem quality, which is biodiversity. This is because international treaties and declarations –

the most important being the *Convention on Biological Diversity* signed by 177 countries (UNEP 1992) – stress the importance of biodiversity for both its intrinsic value and its value in the functioning of the world's ecosystems. Land use is regarded a major threat to genes, species, and ecosystems, which are vital components of biodiversity (Korneck and Sukopp 1988; Perrings 1995; OECD 1996, p. 43).

Part A

Problems, objectives, and foundations

1 The research process in applied sciences

The ultimate goal of applied research is to provide solutions for the practical problems we face. This implies that a prerequisite for this kind of research is to take the practice with its problems into account. By way of contrast, in pure research this has a more optional character. Although solving practical problems is seen as the ultimate goal of applied research, solving research problems is an equally important objective.

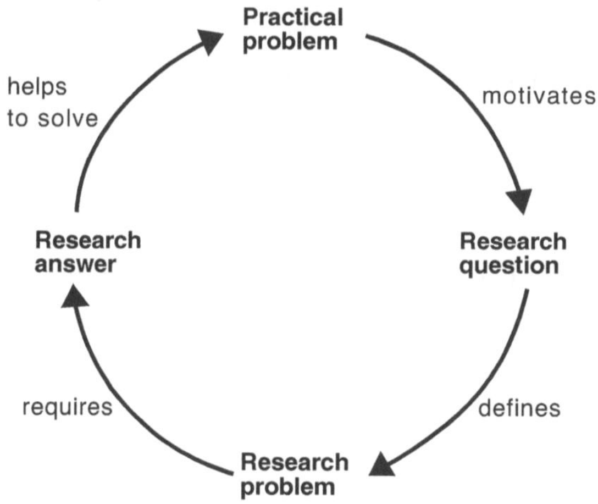

Fig. 1-1: The phases of applied research (adapted from Booth et al. 1995, p. 49).

Dealing with practical problems in science involves another peculiarity of applied research. As practical problems are not organised according to the traditional scientific disciplines, multidisciplinarity is frequently regarded as a requirement, especially with respect to environmental problems (Norgaard 1989; Jaeger and Scheringer 1998).

According to Booth et al. (1995, p. 49) the phases of applied science can be arranged in a cycle (Fig. 1-1). The starting point is a practical problem. This practical problem motivates research questions, which, in turn, define the research problems. Finally, the research answer helps to solve the practical problem.

In order to guide the process of research, several structural elements are introduced to complete the research cycle (Tab. 1-1). At the beginning, the *background information* necessary to understand the practical problem, as well as its embedding in the practice, is given. The *objective* of applied research includes both practical and theoretical components. This *objective* creates a research space and motivates *research questions*. Clarifying the significance of the problems dealt with (both in practice and science) and framing the theoretical and methodological background provide the *embedding* of the research process. A normative embedding might be necessary in addition. In order to solve the research problems it is necessary to apply *methods,* which are rooted in either engineering, natural, or social sciences. One part of the research answer is the *proposed concept* for "solving" the practical problem. *Discussing* the pros and cons of the concept is seen as another, integral part of the research answer. The *application* of the proposed concept in a case study is the last step, which completes the cycle. This step provides the opportunity to evaluate the proposed concept and to develop further improvements. Tab. 1-1 shows an overview of research phases and structural elements of this Ph.D. thesis.

Tab. 1-1: The research phases and their structural elements.

Phase of research	Structural elements	Content
Practical problem	Background information	• Land use initiates environmental problems • Life Cycle Assessment (LCA) is a tool for supporting environmental planning and decisions making • Integration of land use impacts in LCA is necessary
	Practical problem	• Lack of generic information regarding the damage due to land use and of a method for assessing it within the framework of LCA
	Objective	• To propose a method for assessing environmental damage/benefits due to land use • To propose generic information on the damage associated with different land use types
Research question		• How can one best model impacts on ecosystem quality due to land use within LCA?
Research problem	Embedding in research	• Increasing activity in European research institutes as well as in industry e.g., Scandinavian and Canadian wood industry (Stora, Canfor, AssiDomän)
	Embedding in theory	• Life Cycle Assessment, Ecology
	Research problems	• Validity, reliability, uncertainty and lack of empirical information
	Research methods	• Conceptual modelling, quantitative modelling and case study
Research answer	Proposed concept	• Method for assessing land use impacts that takes the attribute *biodiversity* into account
	Application	• Case study

2 The practical problem: Evaluation of land use activities in Life Cycle Assessment

2.1 The decision and its context

Thousands of ecologically based decisions are made annually by policy-makers and regulatory agencies, land- and water-use planners, resource-managers, business and industry, consulting firms and conservation groups. To be useful to decision-makers, ecological information must be both accessible and relevant to their mandates and responsibilities. (Lubchenco 1991, pp. 375)

The intention of my work is to provide such ecological information to decision-makers with respect to the consequences of land use. It is generally possible to distinguish two types of land use decisions. One type of decision about land use is the type made using information about a *specific case of land use* and the site where this case takes place. For example, when alternative locations for an industrial plant or alternative routes for a highway exist and the decision must be made for one of the alternatives. An environmental evaluation of the land use activity associated with the case can be made site-specific with an Environmental Impact Assessment (EIA).

The other type of decision is made considering the environmental impact of a *land use type* without taking into account the specific site of its use. For example, a political decision was made in Switzerland in favour of more organic agriculture and against conventional farming. In such instances, the environmental evaluation of the land use activities which is made cannot be site-specific, but rather, must be generic. However, it may be useful to distinguish different regions in the evaluation.

In LCA the second type of land use decision is more important. The site where a land unit is occupied is generally not known in LCA. This is not a problem of principle, but rather, information costs would rise considerably, if the exact location of every process in a product's life cycle would have to be registered. For example, the raw materials for the concentrate feed that is used to raise livestock in Sweden are produced in Argentina (sunflower), Brazil (soybean), Denmark (sugar beet), France (sunflower, maize), Germany (rape seed), Malaysia (palm oil), Poland (sugar beet), South America (sugar cane), and Sweden (sugar beet) (Cederberg 1998, p. 23). It would be a time-consuming task to specify the exact sites of arable fields in each country and to assess the environmental damage due to biomass production specifically for each site. Therefore, in LCA a more generic assessment of the environmental impact of different types of land use is applied. Of course, when available, case- and site-specific information, can improve the evaluation. It is more likely that such information is available for central processes than for background processes.

2.2 *Framing the practical problem*

Land use and its environmental burdens are seen as the practical problem on a remote level. This is the ultimate problem, and it constitutes the research cycle. It is widely recognised that in Western Europe, but also in other parts of the world, most of the land area available is used rather intensively, initiating environmental problems like overfertilising, load with biocides, erosion, sealing, impact on water cycle, and direct destruction of habitats (European Environment Agency 1998). McCloskey and Spalding (1989) estimate that only 2.8% of the European landmass is still wilderness, compared to 20% in South America. Many governmental and non-governmental organisations are concerned with these problems and attempt to motivate actions that would reduce these environmental burdens.

At the same time Life Cycle Assessment (LCA) is a method that is widely used for backing up environmental planning and decision-making on the level of product systems. It allows decision-makers to compare the environmental damage of product alternatives and to recognise any potential for improving the environmental performance of product systems. The relationship between the method LCA and the objective to improve the performance of product systems is one between means and ends.

The proximate problem is the fact that the impact category *land use* is not addressed in a satisfying manner in LCA. However, land use is assumed to be an important impact category for many product families (e.g., food, pulp, paper, cotton clothing, transportation, energy, and chemical products). This is because at different phases of the product life cycle it is necessary to use land rather intensively. Therefore, assessment of the environmental damage due to land use cannot be excluded from a comprehensive LCA. Furthermore, the incorporation of land use into LCA enables environmental decision-makers to recognise the potential for improving the environmental performance of their product system. This holds especially true for planners and decision-makers who are responsible for a phase of the life cycle which is remote from the main land use activities. For example, decision-makers in the car industry may think that land use is not important for their product, because the land area necessary for assembling cars is not substantial. However, they are overlooking the land area necessary for extracting raw materials, using the car, and disposing the product at the end of its life.

One part of the solution to the proximate problem is to propose a methodological approach. This completes existing LCAs with the element of land use. In addition, it is necessary to quantitatively evaluate the environmental damage and provide, in an accessible form, generic characterisation factors for important land use types. These generic factors are necessary, when decision-makers do not have sufficient resources for performing elaborate case-specific Life Cycle Impact Assessments.

3 Objectives and research questions

Objectives

Development of conceptual model. I conceptualised the relationships between land use activities within the product systems, their impact on the quality of ecosystems, and the values of stakeholders. This provided a framework for the evaluation on the operational level. The difference between environmental effects (i.e., anthropogenic changes in the environment) and damage (negative effects) will be addressed here.

Method for evaluation of land use in LCA. My objective was to develop a method for quantifying the use of land in product systems and the resulting ecological damage within the framework of Life Cycle Assessment (LCA). The proposed method would need to be applicable to both the main land use types (building, mining, traffic, agriculture, and forestry) and the applied management styles within these types (e.g., wood plantations or near-to-nature mixed forest). The main focus of this book is on land use types that are related to agriculture. The differences between local and regional impacts are taken into account.

Generic characterisation factors for assessing specific land use types with respect to biodiversity. The ecological damage for the most important land use types was quantified taking the attribute *biodiversity* into account. The quantification was made with empirical information about plant species diversity. This resulted in a database ready for application, with generic characterisation factors expressing the ecosystem damage potential for species diversity ($EDP^{sp\text{-}div}$) due to land use.

Both the method of characterisation and the generic quantitative information on land use impacts could possibly provide new elements for improving existing Life Cycle Impact Assessment methods. For example, they might be integrated into the Eco-indicator 99

(Goedkoop and Spriensma 1999), the EPS method (Steen and Ryding 1992; Steen 1997; Steen 1999), or the methods proposed in Heijungs (1992) and Kortman et al. (1994).

Application of characterisation factors in a case study. A case study was conducted on agricultural products in order to apply the characterisation factors developed. This allows the environmental impacts of different land use activities in the product life cycle to be assessed.

Research questions

- Which attributes of the ecosystem are valuable? Which types of value provide biodiversity? What is the relation between land use and biodiversity? -> *development of conceptual model*
- How can one assess environmental effects due to land use in LCA? How can one transform observable effects into damages and benefits? -> *method for evaluation of land use in LCA*
- Which specific endpoints should be considered in the operational model? Which indicators for biodiversity are appropriate? What is the validity and reliability of the results? -> *generic characterisation factors*
- Which data are needed in a case study? How are characterisation factors applied in order to calculate the total damage as a result of land use? -> *case study*

4 Disciplinary and methodological foundations

4.1 Life Cycle Assessment

Life Cycle Assessment (LCA) is a tool which takes product life cycles into account when backing environmental decisions and planning in companies, governmental organisations, and non-governmental organisations (Braunschweig and Müller-Wenk 1993; Consoli et al. 1993; Bundesumweltministerium and Umweltbundesamt 1995; Guinée 1995; Kreikebaum 1995; Kytzia 1995; Hanssen et al. 1996; Janzen 1997; Rubik and Teichert 1997). For this, decision-makers need information that is normally beyond their sphere of perception and influence. This is (i) information about environmental interventions (e.g., energy and resource use, emissions) downstream and upstream the product life cycle and (ii) information on potential damage to selected safeguard subjects (resources, human health, and ecosystems) due to environmental interventions. Extended access[1] to information in these two dimensions is seen as essential to improving entire product life cycles environmentally through informed planning and decision. Otherwise, decision-makers miss the chance to explore the potential for improving environmental performance well beyond their immediate sphere of influence. However, the costs for this access to information are relatively high (Schaltegger 1997), despite the fact that background inventory data on generic processes and weighting factors reflecting the environmental burden of processes are provided in databases (e.g., Frischknecht et al. 1996).

Life Cycle Assessment, according to ISO 14040-48, is divided into 4 parts (adapted from Wenzel et al. 1997; Hofstetter 1998):

1 Other environmental instruments, like green labelling based on certification schemes, e.g., FSC-Labels (Forest Stewardship Council) for "green" wood, serve this purpose as well.

1 In *goal definition,* the purpose of the study and the application of its results are defined. The object of study, the functional unit, the system boundaries, the interventions to be included and the scope of the study in relation to the application are settled during *scope definition.*

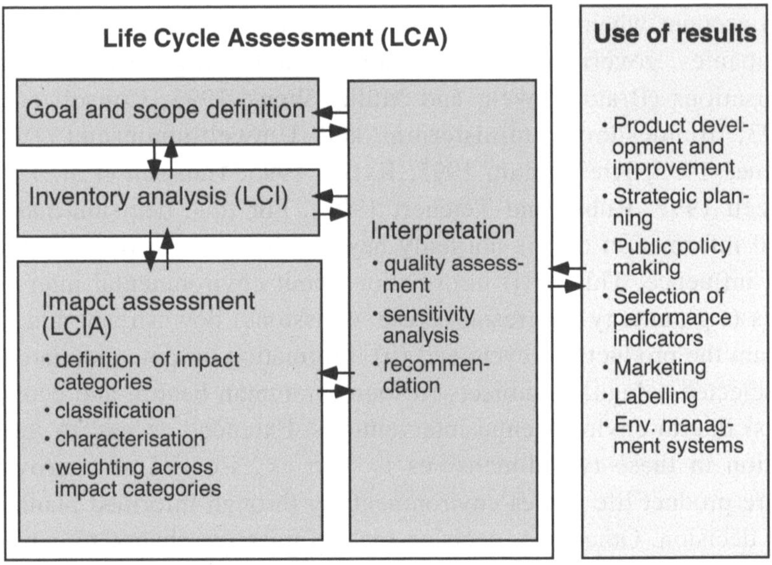

Fig. 4-1: General framework of Life Cycle Assessment (adapted from ISO 1997).

2 In *Life Cycle Inventory (LCI),* environmental data about the product life cycle is collected. Environmental interventions include the amounts of energy, mineral resources, water, biomass, and land which are used in the life cycle. In addition, the amounts of chemicals emitted into air, soil, and water, as well as ionising radiation, and noise released from the product system, are analysed.

3 *Life Cycle Impact Assessment (LCIA)* (Fig. 4-2) aims at evaluating the interventions registered in the inventory (e.g., Steen and Ryding 1992; Goedkoop 1995; Braunschweig et al. 1996; Udo de

Haes 1996; Müller-Wenk 1997; Hofstetter 1998; Udo de Haes et al. 1999). The interventions are assigned to the selected impact categories (classification) with respect to the environmental mechanism at play. This means that different interventions which follow a similar cause-effect path are put together into a single impact category. The potential impact of every intervention within a category is determined individually (characterisation). In a damage-oriented type of LCA, the impact categories are then assigned to safeguard subjects. The contribution of each impact category to the damage of a safeguard subject is quantified based on a modelling approach. Finally the importance of safeguard subjects is weighted using panel methods. In the ISO-framework of LCA, the weighting across different impact categories is an optional element.

4 The *interpretation* phase refers back to all of the other parts of a LCA. Here, the final results are judged with respect to the quality, uncertainty, and goal of the study.

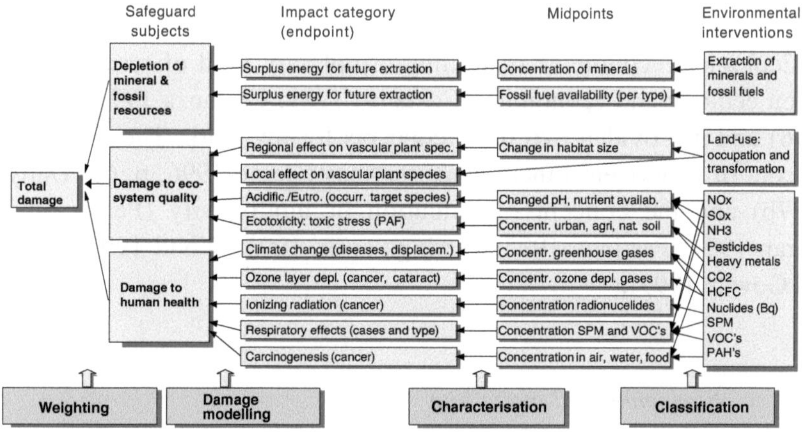

Fig. 4-2: Life Cycle Impact Assessment according to the Eco-indicator 99 (adapted from Goedkoop and Spriensma 1999).

The results of LCA can be useful in very different applications. However, the environmental information coming out of LCA needs to be completed with other information (e.g., information about the financial costs of alternatives) to back up decisions.

4.2 Ecology and ecological economics

Ecosystem theory, which is a component of ecology, provides a framework for the assessment of the land use impacts on biodiversity and ecosystem functions. What is important with respect to my project is the research done on the relationship between biodiversity and ecosystem functions (Ehrlich and Ehrlich 1981; Schulze and Mooney 1994a; Schaefer 1996; Diemer 1997; Finlay 1997), on the anthropogenic causes of biodiversity decline (Fukarek 1980; Ellenberg 1985; Korneck and Sukopp 1988; Ellenberg 1989), on the species-area relationship (Arrhenius 1921; Preston 1962; Connor and McCoy 1979; May et al. 1995), and especially, on impacts of land use on biodiversity (Schumacher 1980; El Titi 1991; Gilbert 1994; Forman 1995; Duelli 1997; Giller 1997).

Ecological economics is an interdisciplinary field of research between ecology and economics (Costanza 1989; Costanza 1991; Bergh 1996) and it provided two basic concepts for my work. These are the concept of ecosystem functions & services (Bergh 1996, p. 62; Daily 1997b) and the concept of valuation of biodiversity (Pearce and Moran 1994; Perrings 1995; Perrings et al. 1995; Wood 1997; Seidl and Gowdy 1999).

4.3 Environmental modelling

Models replicate systems in a simplified manner. A system consists of compartments and relations between these compartments (Bossel

1994; Hoffmann 1997) and is an integer entity; that is, division of the system results in a loss of functionality. The objectives of modelling a system are, first, to support scientific knowledge (e.g., insights in complex system behaviour) and, second, to help solve practical problems (i.e., decision support, pathway analysis, and policy analysis[2]).

According to Costanza et al. (1993) it is possible to distinguish between conceptual models, analytical models, impact-analysis models, and indicator models (see also Bergh 1996, p. 49). These types of models are specified in Tab. 4-1.

The aim of the *conceptual model* in my work is to describe the product system, the ecosystem and values of stakeholders as well as the relationships between these compartments. It is a verbal model which is used to frame the quantitative evaluation of land use impacts. The safeguard subject Ecosystem Quality (EQ) is defined and the attributes of EQ described below.

Tab. 4-1: Types of models (adapted from Costanza et al. 1993)

Type of model	Characteristic	Description	Examples
conceptual models	highly general	• reflect general relationships on a conceptual level; • give up precision and realism; • contain verbal descriptions;	general economic-ecological models
analytical models	highly precise	• aim at high quantitative correspondence between the data and the model; • model small scale and short term changes which results in reduction of complexity and un-	models of energy fluxes and matter cycles in ecosystems (Ellenberg et al. 1986)

2 Pathway analysis helps to answer the question "Which output (e.g., environmental damage) do we get for a defined input into the system (e.g., land area used)?", whereas policy analysis addresses the question "Which input is necessary to get a defined output (policy outcome)?" (in sensu Bossel 1994, p. 248-261).

Type of model	Characteristic	Description	Examples
		certainty; • can have a mechanistic character;	
impact-analysis models	highly realistic	• consider complexity, uncertainty, nonlinearity, and dynamics; • characterised by generality and reduced precision;	simulation of using openly accessible resources based on game theory (Grant and Thompson 1997)
indicator models	Moderate generality and moderate precision	• determine the overall magnitude and direction of change; • being aware the complexity, indicators are used for measuring changes and impacts not directly comprehensible;	environmentally adjusted net national product (green NNP) and models reflecting environmental policy cycles [a]

[a] e.g., Pressure-State-Response Model (OECD 1998), Driving forces-Pressure-State-Impact-Responses Model (Holton-Andersen et al. 1998)

The *indicator model* is used to quantitatively evaluate the environmental interventions due to land use activities. Towards this end, assessment endpoints for the attributes addressed in the conceptual model are defined. Indicators are established in order to quantify impacts on the endpoints. The quantifications in this book are based on publications and databases containing empirical findings for the selected indicators (e.g., data on species number per land use type).

Part B

Ecological evaluation of land use in Life Cycle Assessment

5 Conceptual model: land use, ecosystem, and values

A conceptual model is set up linking land use and ecosystem quality (Fig. 5-1). The objective of this modelling is to provide a framework for the Life Cycle Impact Assessment of land use. Following a top-down modelling approach (Müller-Wenk 1997; Hofstetter 1998), the safeguard subject *ecosystem quality* (EQ) is defined at the beginning. Specific attributes of EQ and corresponding values as stated in stakeholder documents are described. The link between land use and damage to EQ is elaborated upon.

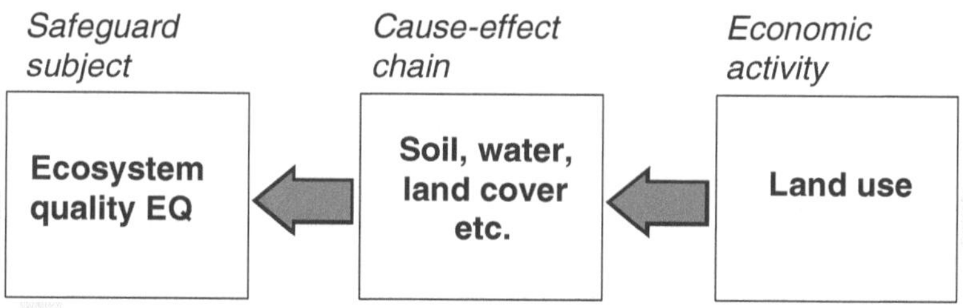

Fig. 5-1: Conceptual model for land use assessment.

5.1 Safeguard subject ecosystem quality (EQ)

To develop the safeguard subject *ecosystem quality* (EQ) requires explicitly considering its basis in natural science as well as society's perception and valuation of ecosystem attributes. Both aspects of EQ are addressed in this chapter.

5.1.1 Ecosystem on the landscape scale

Tansley (1935) shaped the idea of the ecosystem and stressed: "Though the organisms may claim our primary interest, when we are trying to think fundamentally we cannot separate them from their special environment, with which they form one physical system." However, the idea of an ecosystem as a physical, and implicitly a deterministic system, is less prominent in contemporary ecosystem theory: "Ecosystems are not machines" (Müller 1997, p. 156). Instead, emerging properties, complexity and uncertainty are emphasised (Breckling and Müller 1997; Müller 1997).

On a landscape scale, an ecosystem is a heterogeneous system covering the whole area of a region (e.g., the Swiss Alps or Europe). It consists of a pattern of smaller homogeneous ecosystems (Forman 1995, p. 13). These homogenous ecosystems can be forests, agro-ecosystems, urban ecosystems, or lakes. They can vary considerably with respect to their ecological quality and value.

A system's structure, processes, functions, and organisation can be distinguished (Jantsch 1989). Biotic (e.g., bacteria and trees) as well as abiotic parts (e.g., water and soil matrix) are arranged in a *structure*, which can be described with respect to space and time (Schaefer 1992, p. 231). In order to analyse the flow of energy, matter, and information, it is necessary to distinguish between primary producers (plants, algae), consumers (herbivorous, carnivorous animals) and decomposers (bacteria, fungi) (e.g., Ellenberg et al. 1986). The total of various ongoing *processes* together compose and facilitate ecosystem functions (e.g., cycling of nutrients). According to Jantsch (1989, p. 334) *functions* are considered overall characteristics of ongoing processes. System borders can be drawn with respect to these functions.[3]

Another way to look at a landscape ecosystem is to consider the *organisation* of it. The ecosystem can be divided into subsystems, each with its own functionality. Those autonomous subsystems can be individuals, groups of individuals (e.g., functional groups like biomass

[3] Similarly, in LCA the structure of the product system is defined by its function (e.g., packaging) and described in terms of the processes which are necessary to make the function available (e.g., raw material extraction, production of packages, transportation, recycling). In other words, the function (packaging) is the overall result of ongoing processes in the life cycle of a product.

producers or evolutionary groups like species), species communities, or sub-ecosystems. The different levels show a nested, hierarchical organisation (see also O'Neill 1989, p. 143; Müller 1997, p. 150). This means that each level includes its subordinate level; for example, the ecosystem level includes the level of communities.

5.1.2 Attributes of the safeguard subject ecosystem quality

Within the rather general safeguard subject *ecosystem quality* (EQ), more specific attributes can be defined by considering different international treaties and declarations (UNCED 1992a; UNEP 1992; Council of Europe et al. 1996; Held and Kümmerer 1997). These attributes are biodiversity B, ecosystem functions & services F_{eco}, and ecological resources R_{eco} (Fig. 5-2). EQ is considered a function of these attributes:

$$EQ = f(B, F_{eco}, R_{eco})$$

Biodiversity

The concept of diversity comes from information theory and refers to the variability of information contained in a system. In general, diversity is a function of the count (the more elements, the more diversity), the distribution (the more equitable the counts of single elements represented in a large set, the more diversity) and the dissimilarity (the less similarity between elements, the more diversity).

The attribute *biodiversity* is specified as either genetic, species, or ecological diversity (Heywood and Baste 1995, p. 10; di Castri and Younès 1996, pp. 2-5). On the *genetic level,* diversity is divided into the diversity of genotypes within populations or species and the diversity between species. Therefore, rare species, which obviously have a genetic structure which is also rare, contribute more than average to the total genetic diversity.

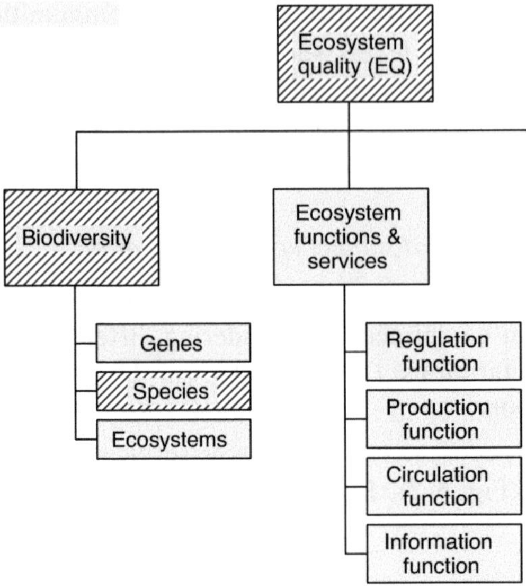

Fig. 5-2: Hierarchy of the attributes covered by the safeguard subject "ecosystem quality" (EQ). The attributes that are addressed explicitly in this book are highlighted in striped boxes.

On the *species level,* species richness, species diversity[4], and species accumulation rate are distinguishable. Species richness indicates the number of species per area or volume of habitat; while diversity in a narrow sense additionally takes the species' abundance (i.e., the number of individuals of each species in the total pool) into account (Schaefer 1992). The diversity increases as the evenness[5] of the spe-

4 Species diversity is defined in a narrow and technical sense with reference to information theory. Biological diversity, in contrast, is meant in a very broad sense and refers to the variability among living organisms in general (see Convention on Biological Diversity).

5 The eveness of a species group increases as the number of individuals per species gets more equal. The highest eveness is reached when each species has the same number of individuals.

cies' abundance increases. As each individual contains a unique genotype, the number of individuals also contributes to genetic diversity.

Ecosystem functions & services

Ecosystem functions refer to the overall characteristic of both locally and regionally ongoing processes in the landscape ecosystem, while *ecosystem services* capture only those functions valuable for human society. Some examples of these functions are regulation of climate, fixation of solar energy, circulation of matter and water, generation of fertile soil, production of food biomass, and regulation of pests (de Groot 1992; Daily 1997a). Recreational and scientific uses of an ecosystem, like one's aesthetic perception of it, can also be regarded as functions. Because these functions require there to be a flow of information between the ecosystem and human beings, they are named information functions (de Groot 1992). Because knowledge about the interdependence of different ecological processes and functions is limited, it is difficult to decide whether a function is of utility to society or not. Therefore, I suggest assigning instrumental values to all ecosystem functions.

Ecological resources

In contrast to ecosystem functions, I define *ecological resources* as stocks of abiotic or biotic components of living nature (cf., Barbier 1995). Some examples of these resources are surface water, fertile soil, food, and non-food biomass, available for use and provide instrumental values to human beings. In contrast to the extractive resource biomass, fertile soil is considered a non-extractive or qualitative resource (cf., Bergh 1996, pp. 62). The difference is in the flow of commodities. Using a qualitative resource does not cause any extraction of commodities. However, in both cases the amount of the resource could be reduced. Resources by definition do not have intrinsic values. Resources of the earth's mineral body (e.g., fossil oil, fossil water, and ore) are not considered to be ecological resources.

Correlation between ecosystem attributes

Each of the three attributes described above accounts for different aspects of the landscape ecosystem from a systemic point of view. Biodiversity denotes the variability of biotic *information*; ecosystem functions & services capture the characteristics of ongoing *processes*; and finally, natural resources are *stocks*. Clearly, the three are not independent. For example, the generation of fertile soil (process) and the amount of soil (stock) are closely linked.

The question is whether changes in biodiversity ΔB, ecosystem functions ΔF_{eco}, and ecological resources ΔR_{eco} correlate with one another strongly:

$$\Delta B \sim \Delta F_{eco} \sim \Delta R_{eco}$$

Biodiversity serves as the basis for sustaining ecosystem functions (Schulze and Mooney 1994a). For the functional relationship between biodiversity (especially species diversity) and ecosystem functioning, several hypotheses have been put forward. The most important are (i) the redundant species hypothesis and (ii) the rivet hypothesis (Lawton 1996, p. 216). According to both hypotheses, ecosystem functioning decreases as species diversity decreases. The difference between the two hypotheses is that different response curves are assumed. According to the redundant species hypothesis, the same degree of species loss results in a moderate loss of ecosystem functioning when many species are left and in a severe loss when only a few species are left. The reasoning behind that hypothesis is that because a specific ecological function (e.g., fixation of nitrogen) can be fulfilled by different species (Lawton 1996) species are redundant. However, the assumption is that all species within one functional group are equally adapted for fulfilling the same specific function (Schulze and Mooney

1994b, pp. 501), but in fact, species within one functional group especially differ in their response to environmental changes. Formerly redundant species might become important for ecosystem functioning, because they are better adapted to new environmental conditions. This point is very much stressed in the rivet hypothesis framed by Ehrlich and Ehrlich (1981). According to this hypothesis, each species in an ecosystem is unique and is necessary to keep the system functioning properly. The analogy is an airplane for which every single rivet is important and the loss of even one rivet (= species) increases the risk of malfunctioning (Lawton 1996).

Based on these hypotheses, several recent studies have found the results for the relationship between species diversity and ecosystem functioning (e.g., Naeem et al. 1996; Hooper and Vitousek 1997; McGrady-Steed et al. 1997; Naeem and Li 1997; Tilman 1997; Tilman et al. 1997; Hector et al. 1999) to be remarkable. However, uncertainty about the empirical evidence still remains. In response to this fact, Schläpfer et al. (1999) conducted an expert survey and asked 39 of the leading scientists doing research in the field of biodiversity about their estimates of the functional relation between ΔB and ΔF_{eco}. It turned out that (i) ecosystem processes and biodiversity are strongly correlated and (ii) ecosystem processes are very important for providing ecosystem services. The results of this survey about the functional relationship support the redundant species hypothesis (see also chapter 7.5.4, pp. 93). Based on these findings it is acceptable to conclude that at least biodiversity and ecosystem functions correlate ($\Delta B \sim \Delta F_{eco}$). However, similar investigations of possible correlations between biodiversity and ecological resources as well as between ecosystem functions and ecological resources, are missing. Thus, making a statement about these correlations is very difficult.

5.1.3 *Typology of environmental values*

Generally, it is possible to distinguish instrumental values from intrinsic values. Instrumental values assume the existence of a relationship between a valued subject and a valuing person. Entities with intrinsic values, in contrast, exist for their own purpose and do not need people to express their values. From a biocentric view plants and animals can be valuable without the existence of people (cf., movement of deep ecology, Sprigge 1997; Lynch and Wells 1998).

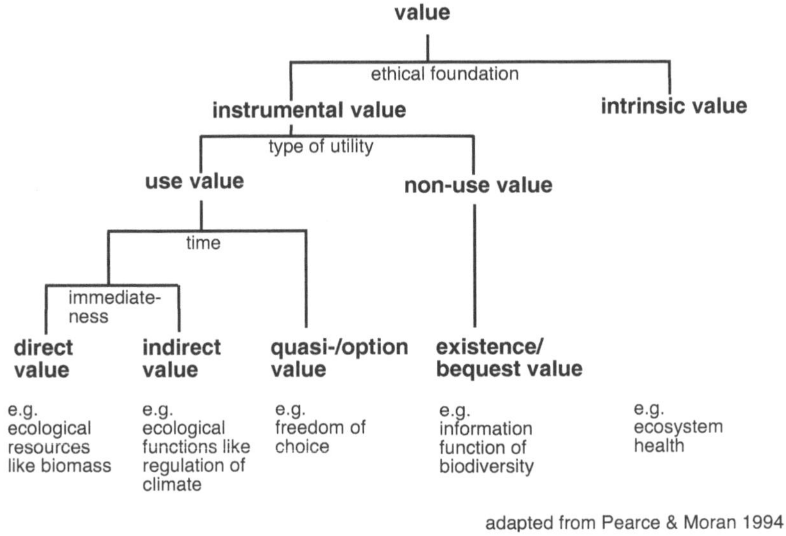

Fig. 5-3: The total value of the ecosystem attributes is composed out of very different values. Pearce and Moran (1994) define the sum of the instrumental values as the total economic value (TEV). Terms written at bifurcations indicate criteria for splitting a value type.

The instrumental value of ecosystem attributes includes very diverse types of values (Fig. 5-3). These use values are direct and indirect use values, option values (i.e., the use values for future generations), quasi-option values (i.e., the value if there is no knowledge about the use values in the future), existence values (for example, knowing that the rainforest exists is of value), and bequest values (the existence value of something - for example, the rain forest - for future generations). Pearce and Moran (1994) call the sum of all those instrumental values of biodiversity the *total economic value* (TEV) of biodiversity. The differences and similarities between existence and intrinsic values are discussed in more detail by Attfield (1998).

Viewing the documents of stakeholders reveal more specific insights into the values of biodiversity. In the *Convention on Biological Diversity,* the attribute is defined as "the variability among living organisms from all sources including inter alia, terrestrial, marine and other aquatic ecosystems, and the ecological complexes of which they are part; this includes diversity within species, between species and of ecosystems."(UNEP 1992, Article 2 Use of Terms). This definition is in line with the ecological meaning described above. The objectives of the Convention "...are the conservation of biological diversity, the sustainable use of its components and the fair and equitable sharing of the benefits arising out of the utilisation of genetic resources,..." (UNEP 1992, Article 1 Objectives). In the Convention, "intrinsic value of biodiversity" and "ecological, genetic, social, economic, scientific, educational, cultural, recreational and aesthetic values" are also distinguished (UNEP 1992, Preamble).

The OECD also provides clear reasons for why biodiversity is important to human society (OECD 1996, p. 7):

- "biodiversity facilitates ecosystem functions that are vital for continued habitability of the planet..."
- "it offers aesthetic, scientific, cultural and other values which are intangible and non-monetary,..."
- "biodiversity is the source of many of the world's products, including foodstuffs, fibres, pharmaceutical products, and chemicals, ..."
- "biodiversity forms the basis for crop and livestock varieties,..."

- "the uniqueness and beauty of diverse ecological systems has value for a wide range of recreational uses and for eco-tourism".

Both documents describe the two basic values of biodiversity: (i) intrinsic, non-tangible values and (ii) values instrumental to human society. The latter is split up into a wide variety of ends for which biodiversity is a means (e.g., eco-tourism, source of foodstuffs, fibres, pharmaceutical products, sustenance of ecosystem functions). In this context, species which are threatened by extinction get a special position, because in this case, intrinsic and instrumental values are irrevocably limited. This special position is demonstrated by the fact that red books exist on the species and habitats which are of concern on supranational, national and regional levels (for example, 34 European countries have elaborated national red lists for threatened species[6]).

For the development of the operational model, it was very difficult to come up with a variety of indicators, each stressing an individual value aspect. Therefore, I focused on the conservation value of biodiversity and its value for ecosystem functioning. The conservation value corresponds to the intrinsic and non-use values, and stresses the concern that species are threatened by extinction. Consequently, the appropriate scale of assessment is the regional one. Its value for ecosystem functioning relates to the use value of biodiversity and stresses its importance for the functioning of ecosystems on the local scale.

5.2 *Land use activities within the product system*

Human activities aimed at making natural resources available are considered land use activities. Depending on the type of resource which is

6 cf., homepage of the European topic centre on nature conservation created by the European Environment Agency (EEA), http://mnhn.fr/ctn/redlist.html, 2.03.99

obtained, the economic process of land use can be classified into three main categories: abiotic resource extraction (mining), biotic resource production (agriculture and forestry), and other surface use (housing, industrial plants, leisure parks, traffic infrastructure, and water reservoirs). In a product system these land use activities can occur in all stages of the life cycle.

A single land use activity might be very simple or it might be a complex composite of different processes. For example, the land use activity *grain farming* is rather complex and requires physically (e.g., ploughing) as well as chemically (e.g., fertilising) based processes. The activities change or maintain the land cover (Turner II and Meyer 1994, p. 5) in terms of its chemical, physical or biological properties (e.g., quantity and composition of surface vegetation, ground water level, and erosion).

5.3 Link between land use and changes in ecosystem quality

In order to quantitatively analyse the ecological consequences of land use activities, the corresponding cause-effect chains are modelled. To do this, both the starting point and the endpoint are defined and the intermediate steps between these points are identified. The endpoints are chosen in accordance with the attributes of the safeguard subject. I have chosen the attribute *biodiversity*, because its decline is of social concern and land use impacts are considered to be a major cause of this decline.

5.3.1 Endpoint definition: Diversity of the regional species-pool

Looking at various definitions of what constitutes biodiversity, it becomes obvious that there is diversity in the possible endpoints to use in the model. However, on the operational level, the attribute biodiversity has to be specified. This is done by selecting appropriate endpoints according to the following requirements (partly adapted from Suter II et al. 1993, p. 22):

- Unambiguous definition
- Susceptibility to all possible land use activities
- Empirical and theoretical foundation
- Accessibility to quantitative assessment

In the quantitative model, I took species diversity and the changes to it which result from land use into account. Genetic diversity is not operational, because there is a severe lack of information about the impact of land use activities on the genetic diversity of populations and species. Ecological diversity is included indirectly, because in the cause-effect chain, the homogenising of habitats (i.e., reduction of ecological diversity) is an intermediate factor that directly contributes to the reduction of species diversity. Requirements for defining endpoints are considered to be as follows:

- *Unambiguous definition:* The specific endpoint considered is the diversity of the regional species-pool. The regional species-pool is generally defined as "...the set of species occurring in a certain region..." whereas the "...region is a reasonably large area of the Earth's surface, with a more or less uniform physiography and climate,..." (Zobel 1997, p. 267). In a broad sense the diversity of the regional species-pool can be described in terms of species richness (i.e., number of species), species accumulation rates (i.e., rate of species increase as area sampled increases), species abundance (i.e., number of individuals per species) and species eveness (i.e., differences in the number of individuals per species). Thereby, all the species and those that are threatened can be distinguished. A similar endpoint, the "probability of occurrence of species", was introduced by Wiertz van Dijk (1992) cited in Goedkoop (1998).

- *Susceptibility to all possible land use activities:* As the species-pool covers a region, all land use activities within this region obviously influence the occurrence of species in the pool. It is assumed that land use activities are an important agent, which has impact on the species-pool.
- *Empirical and theoretical foundation:* The species-pool concept and the mechanisms influencing the species-pool (described by the island theory and mosaic concept) are backed up both theoretically and empirically (e.g., Arrhenius 1921; Preston 1962; Connor and McCoy 1979; Wisheu and Keddy 1996; Caley 1997; Morton and Law 1997).
- *Accessibility to quantitative assessment:* Obviously, it is not possible to assess the potential impact taking all species in the pool into account. Therefore, I have chosen the vascular plant species as a proxy for the total species number. One reason for this choice is the existence of reliable data for a wide variety of land use types. Vascular plant species is also an important species group in ecosystems and its diversity correlates highly with other species groups' diversity (Duelli and Obrist 1998).

5.3.2 Cause-effect chain between land use and the chosen endpoint

There are many factors which shape the diversity of a regional species-pool. These factors are natural conditions, such as climate, solar radiation, water conditions, and soil conditions (e.g., Wohlgemuth 1998). The region's evolutionary history is also important. In addition anthropogenic impacts, whether unintentional (e.g., acidification) or intentional (e.g., use of pesticides), also influence the species-pool.

In Europe, land use is a factor which has had historically a large impact on regional biodiversity and still has today. Fukarek (1980) gives an idea about the development of species richness between 5000 BC and 1975 in Western Europe. Basically there was an increase in species richness up to sometime between 1700 to 1850 and a decline in species richness after that time.

Fig. 5-4: Rather intensive land use about 600 years ago. (Limbourgs, 1412-16, from the series "Les très riches heures du Duc de Berry", left picture: March, and right picture: November, source: Web Museum Paris www.oir.ucf.edu/wm/rh/, 4.2.00)

The increase occurred because almost the entire surface of Europe was converted from its pristine stage into a cultivated one. "Human activities turn natural landscapes into mosaics of croplands, forests, and abandoned areas in different stages of succession." (Lubchenco 1991, p. 392). This land use mosaic was beneficial for regional species diversity, because habitat conditions became more diverse. Although agricultural ecosystems and forests were used rather intensively – as shown in the paintings of the Limbourg brothers 600 years ago (Fig. 5-1) – those forms of human land use in Western Europe enhanced aspects of biodiversity and sometimes still do (e.g., dry grassland). The decline was also partly due to land use. At this time the intensification of agriculture was enormous. In addition, infrastructure increased very much.

In general, the occurrence of land cover types, the pattern of the land cover mosaic, chemical and physical soil conditions, as well as water availability are important midpoints in the cause-effect chain that links land use activities to effects on the species-pool.

The pattern of the land cover mosaic is assumed to shape the diversity of the regional species-pool (Forman 1995; Duelli 1997). The land mosaic consists of patches of different land cover types, which may differ in patch size (e.g., the size of a field) and diversity of species. Therefore, a high proportion of species rich land cover types leads to high regional biodiversity and vice versa (i.e., interior diversity), but the spatial and temporal arrangement of existing land cover types is also an important factor determining regional species diversity. The spatial arrangement determines the total length of edge habitats (e.g., forest-meadow habitat), which serve as habitats for many species (i.e., edge diversity). There is evidence that spatial and temporal homogenising of habitats on the landscape scale causes species decline.

The land use interventions which are most common are desiccation/irrigation, physical soil treatment, and fertilisation. These interventions alter chemical and physical soil conditions and water availability. Intensification of agriculture has mainly resulted in homogenous conditions in terms of the medium availability of water and the high availability of nutrients. This change in abiotic habitat conditions is one reason for the species decline observable in Europe in the last 50 years (Hampicke 1991, pp. 254-259), because many plant species are adapted to low availability of nutrients and extreme water conditions (i.e., either dry or wet).

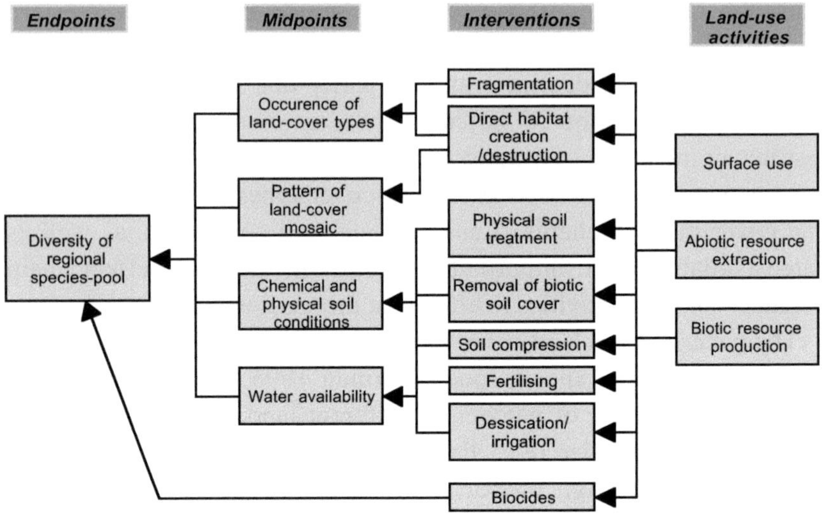

Fig. 5-5: The cause-effect chain with respect to land use. Land use activities, which are economic processes in the product system, are linked with interventions, midpoints, and effects on the endpoint "diversity of regional species-pool" in the ecological system.

However, the model's accuracy strongly depends on one's knowledge of the cause-effect chain. Intermediate steps frequently are not known in detail, and even when they are known, they lack quantitative information. In order to circumvent this fact, I used empirical studies on different land use types and on the species set each type holds. If, for example, 50 species are found on intensively managed fields and 150 on organically managed fields, I assume the increased number on the organically managed fields to be due to the reduced application of biocides and fertiliser, although I am not able to separately assess these interventions.

6 The impact category *land use*

In LCA, an impact category includes impacts which are similar in their environmental mechanism (e.g., chemicals which deplete the ozone layer are grouped together in one category). In this respect, land use is considered an impact category (Finnveden 1996; Heijungs et al. 1997; Udo de Haes et al. 1999). Nevertheless, the cause-effect path for different land use activities can be very variable. It might also be confusing that land use, at the same time, is an economic process in the product life cycle, an environmental intervention, and an impact category.

In this chapter, some topics pertaining to the impact category *land use* are elaborated. These are the temporal and spatial scale of land use effects, the determination of the assessment perimeter, the reference state, and the multifunctionality of land.

6.1 Temporal and spatial scale of environmental effects

Temporal effect

Land use activities can be arranged according to their temporal phase of use. Frequently, four types of activities are distinguished in LCA (cf., Blonk et al. 1997; Baitz et al. 1998, p. 8; Lindeijer et al. 1998, pp. 18-22).

- In order to fulfil a specific economic purpose, it might be necessary to transform the initial land cover type into a desired land cover type. This phase is called *land transformation*. The initial land cover type can either be a pristine one or one which has already been transformed to fulfil previous economic purposes.
- The next phase is *land occupation*. In this case, a certain area of land is occupied for a certain length of time in order to fulfil

a specific economic purpose. During the occupation, the ecosystem quality is constant or might slowly decrease.

- *Land restoration/recovery* takes place after land occupation has stopped. This restoration might be supported by human activity or might occur spontaneously. Restoration is considered to be a special type of transformation. A peculiarity is that restoration times can be very long (e.g., up to 10'000 years for peatbogs).
- The phase *land abandonment* can be a part of the land use activity, at least for those with a defined end of use phase (e.g., mining). The difference from land restoration is that ecosystem quality is assumed to remain constant.

Normally ecosystem quality varies throughout the different phases. The temporal effect on the environment of a complete sequence of land use is indicated by the area shaded grey in Fig. 6-1. The proportions of the parts might change considerably with respect to a specific case, e.g., the restoration phase can be much longer than the occupation phase with the consequence that the restoration effect is much more important than the occupation effect.

Edge effect

The edge effect occurs as a result of different land use types adjoining one another in the landscape. At the border between two land use types, a change in ecosystem quality EQ occurs. For specific land use types (e.g., the border between agriculture and forest) the ecosystem quality of the edge might be even higher than those of pure land use types (B' in Fig. 6-2).

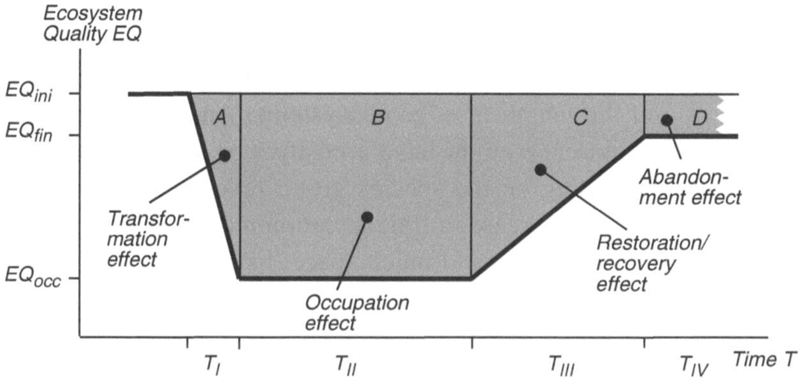

Fig. 6-1: Temporal effect perimeter of land use activities (adapted from Baitz et al. 1998; Lindeijer et al. 2001). The effect of land use is indicated by the area shaded grey and equals the product of time and change in ecosystem quality.

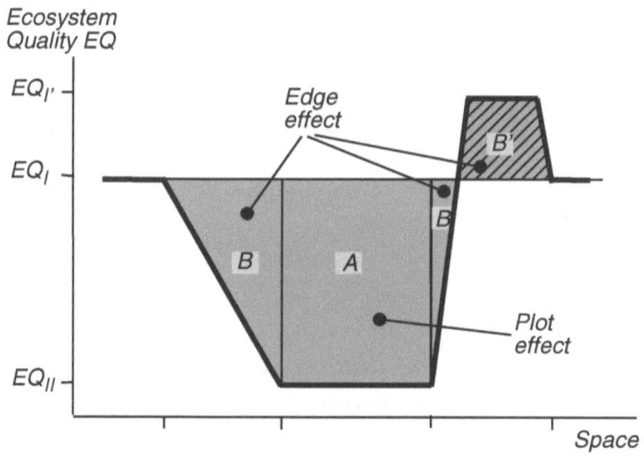

Fig. 6-2: Effects of land use activities in the landscape. The plot effect (A) results from the part of land, which is occupied. Edge effects (B) are a result of the gradient in ecosystem quality of adjoining land use types.

Fragmentation effect

The patchiness of a land cover mosaic has a large effect on the number of plots and the length of edge ecosystems. An increase in fragmentation of a landscape might have a positive or negative effect on biodiversity, depending on the species group investigated. For large animals, like deer, an increase in fragmentation may be negative while for species which don't demand much space, like beetles, this can be positive.

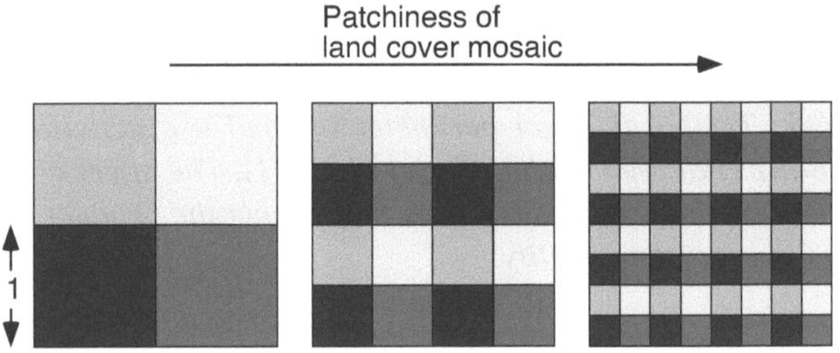

Number of plots	4	16	64
Size of plots	1	0.25	0.0625
Ecosystem fragmentation	low	medium	high
Length of edge ecosystem	low	medium	high

Fig. 6-3: The patchiness of the land cover mosaic and the influence this patchiness has on ecosystem properties.

Local and regional effect

The difference between local and regional effects can best be explained by giving an example. Consider two regions, each completely composed of two land use types: in region *Malus* a large proportion is covered with land use type *m* and a small proportion with land use type *b*. Region *Bonus*, in contrast, has a rather small proportion of land use type *m* and a large proportion of land use type *b*. Consider that the ecosystem quality EQ of land use type *m* is lower than that of land use type *b*:

$$EQ_{land\_use\_m} < EQ_{land\_use\_b} \ .$$

That means the average ecosystem quality of region *Malus* is lower than that of region *Bonus*:

$$EQ_{region\_Malus} < EQ_{region\_Bonus} \ .$$

If, on the local scale, one unit of land use type *b* were transformed into land use type *m*, then the damage on the local scale would be equal in both regions. The amount of damage would be equal to the difference between the levels of quality of *m* and *b* multiplied by the area transformed. Looking at the whole regions, however, this local transformation could be considered to be more detrimental in region *Malus* than in region *Bonus*. This is because of the beneficial land use type *b* already being scarce in region *Malus*. Therefore, the loss of a unit of *b* is more detrimental to this region than it is to region *Bonus*. Technically speaking, the regional effect of land occupation or transformation changes as the average quality of a region's ecosystem changes.

6.2 *Marginal and average approach*

In order to perform local and regional damage/benefits analysis, it is important to decide on the type of assessment approach to apply. A marginal approach can be generally distinguished from an average one (Heijungs 1998).

Both approaches can be adapted to land use. The idea behind the marginal approach is that a region with a mixture of land use types is given as the background situation and small changes have to be assessed considering this situation (Fig. 6-4a). The average approach can be used for the assessment of considerable changes in the proportion of land use types (Fig. 6-4b). Here the changes in ecosystem quality are distributed equally across the percentage change of the land use types. The assumption is that for each land use type, a damage curve can be constructed and that the sum of the individual curves gives the total damage curve for a mixture of land use types.

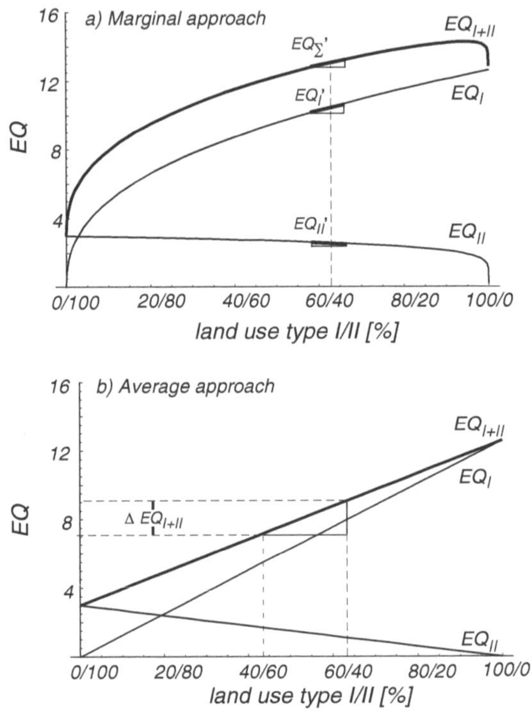

Fig. 6-4: Average and marginal approaches to damage/benefits analysis, where EQ is ecosystem quality, $EQ_{I\ or\ II}$ ecosystem quality of land use type I or II, and EQ_{I+II} ecosystem quality of a specific mix of land use types I and II.

In an exemplary region, *land use type I* occupies 60% of the area and *land use type II* 40%. A rather tiny land transformation might be one in which 1% of type I changes into *type II*. In the new mixture, 59%

would then be composed of *type I* and 41% composed of *type II*. The resulting changes to regional scale ecosystem quality would be best assessed using a marginal approach, because the intervention of 1% is assessed with respect to the background situation in that region. On the plot itself where the land use was formerly *type I* and now is *type II*, 100% of that local area is affected by land occupation. In this case an average approach applies.

In both cases the damage can be assessed by taking the first derivative of the damage curves (Fig. 6-5). In an average case, damage is independent of the mixture of land use types and, therefore, remains constant. For a marginal case, taking the first derivative results in a curve which shows dramatic rates of changes at the extremes of the x-axis. Please note that the first derivative is sensitive to the direction of change between *type I* and *type II*. An increase in *type I* is positive up to a value of about 90%. Any further increase is negative, because the drop in EQ of *type II* is very sharp.

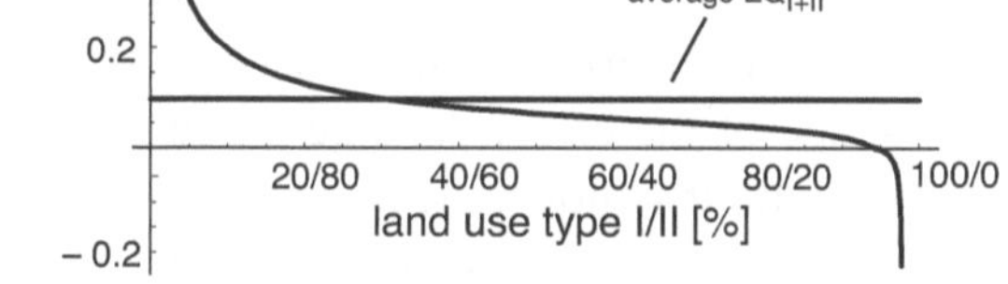

Fig. 6-5: First derivative of the total damage curves (average and marginal approach) which represent a mix of land use type I and II.

6.3 Assessment perimeter

The transformation of land on a local scale is by definition always a complete one. One hundred percent of one land use type is transformed into another one. On a regional scale, in contrast, only a part of one land use type is changed into another one. Whether this percentage is rather large or very tiny depends largely on the perimeter of assessment.

An important question is whether the choice of assessment perimeter has an effect on the result of the damage/benefits analysis or not. To answer this I provide the following hypothetical example. Initially a region is occupied by only one land use type (*land use type*$_{initial}$). This type shows a constant level of ecosystem quality of 0.5. Then a land transformation from *land use type*$_{initial}$ into *land use type*$_{occ}$ of approximately 9 km^2 occurs. The local depression in the "quality" surface is down to 0 (Fig. 6-6). The volume of this depression represents the effect of the transformation (disregarding any restoration effects).

Possible implications of now changing the assessment perimeter include the following:

- The local loss of EQ is constant for graphs *a* to *d* and is equivalent to the volume of the dent in the surface.
- The regional loss of EQ calculated with the *marginal approach* remains constant for graphs *a* to *d*, because the background situation is considered to be constant. However this only applies if the shape of the effect/damage curve remains the same regardless of scale. It is not sensible anyway to apply the marginal approach in cases *a* and *b*, because a large proportion of the total area would be affected. The marginal loss of EQ changes, when the effect/damage curves are different on different spatial scales.

- The regional loss of EQ decreases from graphs *a* to *d* when calculated with the *average approach*, because the proportions of the affected area are different.
- The background situation could change, because the area used to calculate the distribution of land use types increases from *a* to *d*. However, the background situation in the example is constant.

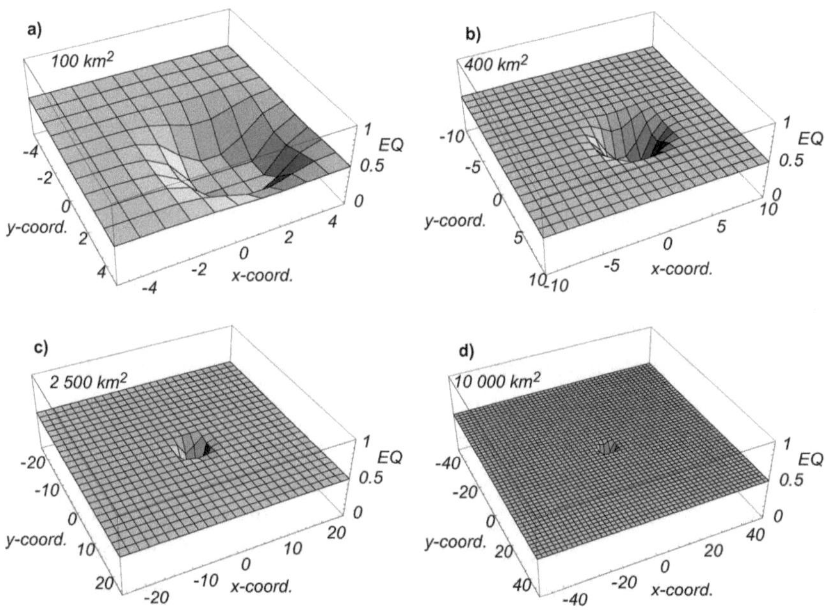

Fig. 6-6: Effect of the assessment perimeter on the calculated loss of ecosystem quality (EQ). A land transformation of ca. 9 km² (1 square equals 1 km²) results in a local drop in EQ. Because of the edge effect, the transition in EQ is smooth.

The conclusion is that when one applies a marginal approach, the results one gets on the regional scale are independent of the assessment perimeter. The underlying assumption, however, is that both the background situation and the shape of the effect/damage curves are constant for different spatial scales.

6.4 Reference state

References are needed to benchmark values of ecosystem quality EQ for different land use types. Possible references for a single land use type in sequence with other types are

- the initial EQ just before the transformation,
- the final EQ after the restoration,
- the best of class (EQ_{max}) land use type,
- the regional reference EQ_{region}, e.g., the mean species richness in the region,
- the historical reference, e.g., the species richness of the region in 1850, and
- a hypothetical reference, like the potential natural vegetation.

It is generally possible to make a distinction between two types of references. First, there are references which are needed for dividing absolute figures (e.g., species number per area) to get a relative measure. Second, there are references which can be used to trace back changes in time resulting from land transformation. Here the choice has a large impact on the results. In the first case the consequences of using the different choices of a reference might be less severe, because the point of the reference is simply to make the types of land use comparable within its impact category using a relative scale. However, for a damage oriented LCA, the choice of reference might have large consequences, since the potential to damage the safeguard subject *ecosystem quality* is assessed across different impact categories, and *land use* is one of them.

6.5 *Multifunctionality of land use*

One problem of the Life Cycle Impact Assessment is that land use can be multifunctional. The objective of modern agriculture is not only to produce food, but also to protect the landscape and biodiversity and to contribute to human recreation. Biodiversity could be considered a by-product of agriculture. Then in the impact assessment only a part of the area used could be attributed to biomass production, the rest to biodiversity "production". However, this is not appropriate for a damage oriented LCA, because biodiversity is already used as an endpoint of impact assessment. One exception might be when biodiversity is defined as the functional unit of two processes or products to compare. A parallel case is the assessment of gas purification systems, where a volume unit of clean air is defined as a functional unit.

6.6 *Existing proposals for land use impacts in LCA*

To compare the different proposals of how to assess land use within the framework of LCA, the conceptual level must be distinguished from the operational level. The conceptual level takes into account specific attributes of the safeguard subject (or area of protection) as well as the safeguard subject itself. Safeguard subjects are not always stated explicitly by the authors. I also tried to assess whether the type of value the authors look at is instrumental or intrinsic. On the operational level, I looked at indicators for the specific attributes, the type of assessment, and the main area of application. Other reviews of existing proposals are found in Finnveden (1996), Ekvall (1997) and Heijungs (1997).

Tab. 6-1: Conceptual level of proposals for land use in LCA.

Conceptual level	Safeguard subject stated by the authors	Attributes of safeguard subjects	Type of value[(b)]
(Anonymus 1998)	Robust and functioning ecosystems	1: Biological diversity; 2: Maintenance and enhancement of forest ecosystem condition and productivity; 3: Conservation of soil and water resources	1, 2, 3: Intrinsic & instrumental
(Audsley 1996, p. 85)	Soil quality	Soil fertility ("ability of the soil to grow a good crop")	Instrumental
(Baitz et al. 1998; Baitz et al. 2000)	Quality of the natural landscape [(a)]	Eight functions of landscape: erosion resistance, filtering and buffering properties, decomposition of organic matter, influence on ground water, flood regulation, immissions protection, biological stability, biomass production	Intrinsic & instrumental
(Cowell 1998)	Ecosystems?	Biodiversity	Intrinsic & instrumental
(Ekvall 1997, pp. 20)	Environment	Biological diversity; productivity of forests; cultural, aesthetic, and recreational values; demand for the resource productive land area	Intrinsic & instrumental
(Hofstetter 1994, pp. 44; Frischknecht et al. 1996)	Environmental health	Naturalness of land	Intrinsic
(Giegrich and Sturm 1996)	Naturalness of ecosystems	1: Naturalness of soils; 2: Naturalness of forest vegetation ; 3: Naturalness of conditions for ecosystem development	1,2,3: Intrinsic
(Goedkoop et al. 1998; Goedkoop and Spriensma 1999)	Ecosystem quality	Biodiversity	Intrinsic
(Heijungs et al. 1997, pp. 28)	Environment	1: Reduced availability of land for securing human welfare and well being; 2: Loss of biodiversity; 3: Damage to life support	1: Instrumental, 2: Intrinsic, 3: ?
(Köllner 2000)	Ecosystem quality	1: Biodiversity; 2: Ecosystem functions; 3: Natural resources	1,2,3: Instrumental

Conceptual level	Safeguard subject stated by the authors	Attributes of safeguard subjects	Type of value[(b)]
(Lindeijer et al. 1998)	A: Ecosystem quality; B: Resources	A1: Biodiversity; A2: Life support; B: Reduced availability of land	A1, 2: Intrinsic, B: Instrumental
(Müller-Wenk 1998)	A: Human health; B: Ecological health; C: Resources	A: Increased probability of floods and landslides arising from land use resulting in human victims; B: Biodiversity; C: Quality of soil and quantity of water	A: Intrinsic, B: Intrinsic, C: Instrumental
(Schweinle 1998)	Environment?	1: Soil; 2: Water; 3: Vegetation	?
(Swan 1998)	Potential bioquality	1: Bioproductivity; 2: Biodiversity	1,2: Intrinsic
(Udo de Haes et al. 1999)	A: Natural resources; B: Natural environment	A: Increase in land competition; B1: Degradation of life support functions; B2: Biodiversity degradation	A, B1, B2: Instrumental and intrinsic

[(a)] *The term in the German publication is "Qualität des Naturraums".*
[(b)] *This requires interpretation in some cases; therefore, the results are disputable.*

During the time of the project, many suggestions to include land use in LCA were made. Many safeguard subjects and a variety of their attributes are supposed to be impacted by land use. Three basic types of safeguard subjects are mentioned: human health, ecosystems, and resources. The very fundamental safeguard subject *environment* is also given. Attributes of ecosystems are considered to be biodiversity, ecosystem/life support functions, productivity, and naturalness. Attributes of resources are land competition, quality and quantity of soil, and quantity of water. In addition landscape aesthetics is mentioned in Blonk and Lindeijer (1997).

Tab. 6-2: The operational level of the proposals listed in Tab. 6-2.

Operational level	Proposed indicators for operational models	Type of assessment	Perimeter	Main application
(Anonymus 1998)	Management practices regarding 1: size, fragmentation, and quality of habitats; also conservation of forest genetic material; 2: protection of ecological stress/ productivity resilience; application of biocides; 3: soil degradation, prevention of chemical leakage; impact on hydrological processes	Forest management practices are assessed on an ordinal scale with 5 categories (best practice to worst practice)	Local	Forestry
(Audsley 1996)	–	–	–	Agriculture
(Baitz et al. 1998)	For each of the eight functions, indicators are described in detail	Performance of functions are assessed on an ordinal scale (5 categories) for specific land use types; total performance of functions is integrated over time [(a)]	Local	Land cover types with defined end of use (e.g., mining or landfill)
(Cowell 1998)	Contribution of land use type to global biodiversity, number of rare species, number of species, Net Primary Productivity NPP	Differences are assessed on a cardinal scale	Global, regional, local	All types of land use
(Ekvall 1997)	–	–	–	Forestry
(Hofstetter 1994, pp. 44; Frischknecht et al. 1996)	Recultivation time	Land use types are assessed on an ordinal scale (4 categories) according to naturalness; recultivation times are assessed for transformations from less natural types to more	Local	Mining, energy systems, waste management

Operational level	Proposed indicators for operational models	Type of assessment	Perimeter	Main application
		natural types		
(Giegrich and Sturm 1996)	1: Intensity of soil use, fragmentation, intensity of chemical impacts, undisturbed water cycle; 2: Diversity of trees, naturalness of vegetation, structural diversity of forest, amount of dead trees; 3: Spontaneity of forest and vegetation development	Forest types are assessed on an ordinal scale (5 categories) according to naturalness of attributes	Local	Forestry
(Goedkoop et al. 1998; Goedkoop and Spriensma 1999)	Potentially disappeared fraction of vascular plant species	Changes in percentage of species which result from land use are assessed on a cardinal scale	Local and regional	All types of land use
(Heijungs et al. 1997, pp. 28)	–	–	–	–
(Köllner 2000)	Local: Relative number of species per type of land use; Regional: Marginal changes in species number for a given fraction of high-intensity land use in the region	Relative changes are assessed on a cardinal scale for different land use types	Local and regional	All types of land use
(Lindeijer et al. 1998)	A1: Species accumulation rate per land use type; A2: Free net primary biomass productivity per land use type	Changes are assessed on a cardinal scale for different ecosystem types worldwide	Local	Abiotic resource extraction
(Müller-Wenk 1998)	B: Percentage of threatened plant species within a region	The increase in number of threatened plant species, which occurs as land use becomes increasingly	Regional	All types grouped into high-intensity and low-

Operational level	Proposed indicators for operational models	Type of assessment	Perimeter	Main application
		intensive within a region, is assessed on a cardinal scale		intensity land use
(Schweinle 1998)	1: Soil compaction, ground sealing, soil erosion, soil chemistry; 2: water balance, water quality; 3: Number of species separated into different groups (tree, shrub, and herb species)	Changes are assessed on a cardinal scale	Local	All types of land use
(Swan 1998)	–	–	Local	Forestry, Agriculture
(Udo de Haes et al. 1999)	A: Area of exclusive land use for a given period of time; resource quality of land; B1: Vegetation cover, standing crop, productivity of natural vegetation, soil permeability; B2: Loss of vascular plant species per area	–	–	All types of land use

[a] The assumption made is that the quality of the natural landscape is highest in the transformation phase at the beginning, is lowest during the occupation phase and recovers gradually after stopping the occupation.

On an operational level, it is possible to distinguish different groups of indicators, which relate to specific safeguard subjects or attributes (Tab. 6-2).

- Some of the indicators are based on *naturalness* including the time necessary to regenerate after different types of use (e.g., Knoepfel 1995; Frischknecht et al. 1996; Giegrich and Sturm

1996). These rest mainly on Blume and Sukopp's (1976) Hemerobien Concept and are, interestingly, the early proposals.

- Some indicators for *biodiversity* are the number/percentage of all the species, the number/percentage of threatened species (e.g., Cowell 1998; Müller-Wenk 1998; Goedkoop and Spriensma 1999), and the species accumulation rate (Lindeijer et al. 1998). Structural diversity in the case of forests (Giegrich and Sturm 1996) and the sizes of scarce and surplus habitats are also mentioned (Müller-Wenk 1997).
- A number of structural and functional indicators are proposed for *ecosystem/life support functions* (e.g., Blonk et al. 1997; Baitz et al. 1998; Udo de Haes et al. 1999). Some examples of these are vegetation cover, productivity of natural vegetation, and energy balance. Lindeijer et al. (1998) propose the free Net Primary Production (fNPP).
- Some indicators relate to *resources*. These are especially those for land competition (the area of land used) and those for soil quality, soil compaction, erosion, and chemistry) (Schweinle 1998).

Most of the methods are based on an assessment of different land use types. These types are ranked either according to their naturalness on an ordinal scale or according to other indicators on a cardinal scales. The source of information used for the assessment is either literature or field surveys.

In Anonymus (1998) a method for assessing the sustainability of different forest management styles is proposed. This method is also based on an indicator approach. However, in contrast to the other methods, the impact is assessed with a questionnaire. This indirect procedure is adapted from the methods used to assess the quality and continuous improvement of management systems.

7 Operational model for assessing impacts on biodiversity

7.1 Synopsis of the model

This chapter contains the development of characterisation factors for different types of land use. These factors allow land use types to be ranked ecologically according to their potential damage to the *diversity of the regional species-pool* (or *species diversity* as an abbreviated form). In Life Cycle Impact Assessment, these factors are used to "...express the contribution of a unit environmental intervention ... to the chosen impact category..." (Heijungs and Hofstetter 1996, p. 32). For the impact category *land use,* the intervention in its most general form is the area multiplied by the period of time for which the land is used. The unit of this surfacetime is hectare multiplied by year. The total damage is then the surfacetime multiplied by the dimensionless characterisation factor:

$$Damage = Characterisation\ factor * Area * Time \qquad [1]$$

In general, the total characterisation factor can be divided into a local factor and a regional factor (Fig. 7-1). The intention of the factor for the local scale is to capture the value of biodiversity for ecosystem functions. The intention of the regional factor is to reflect the conservation value of biodiversity. One reason for this is that the conservation status of a species can only be assessed in the context of a region. Its value for ecosystem functioning, in contrast, can better be assessed in reference to the local scale. Unambiguous assignment of a type of value to an individual scale, however, is not possible.

To develop both characterisation factors, I distinguished between the effect analysis and the damage/benefits analysis. An *effect analysis* is based on empirical information and the results are rather descriptive. On this level, endpoints and indicators for assessment were defined. The effect analysis was performed for the local scale and the regional scale. For the local scale I made a meta-analysis about the species richness of different land use types. In addition I assessed potential restoration times for some important land use types. On the regional scale, I used a regression analysis to capture the relation between the intensity of land use and the corresponding species potentially lost since ~1850.

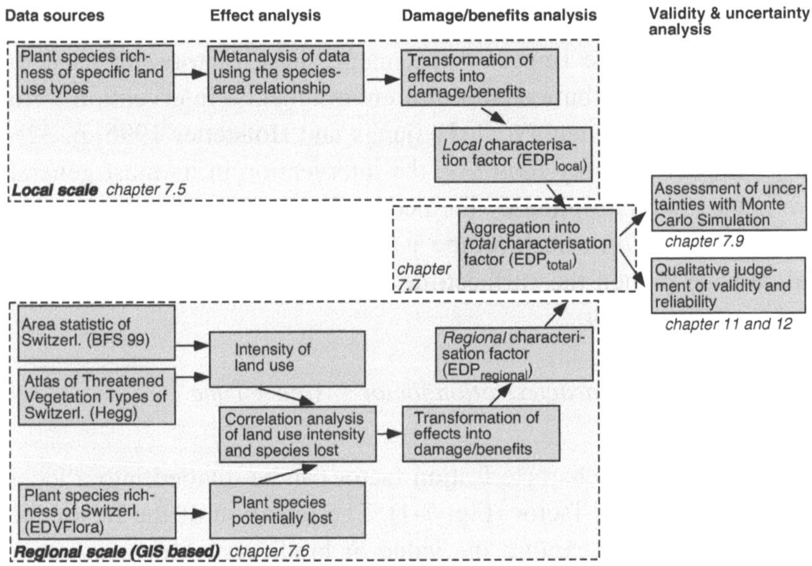

Fig. 7-1: Structure of the model to derive characterisation factors.

The objective of the damage/benefits analysis is to construct characterisation factors, which describe a specific land use type's damaging or beneficial effect on ecosystem quality. This includes transformation of the descriptive results from the effect analysis into damage/benefits, choice of reference, choice of index formation, and defining weighting factors between local and regional damages. For the total characterisation factor, validity, reliability, and uncertainty were assessed.

Basic principles at a glance

To develop characterisation factors, I considered the following basic principles:

- Land occupation and land transformation are regarded as basic types of interventions. *Land occupation* is considered a continuous intervention, which prevents spontaneous land transformation. *Land transformation* means the change from one land use type to another, whether as a result of human activity or not. Restoration of a plot of land after occupation is regarded as a special kind of transformation.
- The assessment of land use is *not site-specific*. That is, it takes information on the land use type, management style, and biogeographical region into account, but not the exact geographical site.
- The method is applicable *across all land use types*.
- The specific endpoint for the effect analysis is the *diversity of the regional species-pool*. That is, the endpoint considered is on the ecological level and not merely on the level of abiotic impacts (e.g., nitrogen concentration in soils).
- The *number of species missing* on the plot in comparison to a reference is the indicator for impacts on the local diversity of the species-pool.
- The *number of species potentially lost since ~1850* in the region is taken as the indicator for impacts on the species-pool on the regional scale.
- *Vascular plant species richness* serves as a proxy for the total species richness. All species are equally weighted, however all the species and those that are threatened are distinguished at the regional scale.

7.2 Types of values taken into account

Biodiversity or, more specifically species diversity provides two basic types of values in the context of ecosystem quality: the value for ecosystem functioning and the conservation value (see also chapter 5.1.3, pp. 32). Changes in species richness due to land use are considered either damages or benefits, because these values are affected.

I consider a change in local species richness (ΔS_{local}) to be an indicator of change in the *value for ecosystem functioning* ($\Delta V_{eco.func.}$):

$$\Delta S_{local} \xrightarrow{indicates} \Delta V_{eco.func.} \qquad [2]$$

This is because species support ecosystem functions (Schläpfer and Schmid 1999) on the local scale. An example of an ecosystem function is the generation of fertile soil, which is supported by a diversity of soil species, like earthworms. In this case it is relatively unimportant per se whether the species are common or rare on a regional scale. What is more important is the widespread occurrence of common species in order to fulfil all of the functions in the ecosystem. For the functional performance, it is merely important that the species are present on the local plot. However, there are keystone species, like beaver, which are at the same time important for ecosystem functioning and threatened on the regional scale.

The *conservation value* of species, however, can best be assessed on the regional scale, because for this information about the degree of regional rarity is necessary. In addition, the degree of threat to a single species and thus its conservation value can be assessed by monitoring the development of its occurrence over time in a region. Therefore, a change in species richness on the regional scale ($\Delta S_{regional}$) over time indicates a change in conservation value $\Delta V_{conservation}$ in that region:

$$\Delta S_{regional} \xrightarrow{indicates} \Delta V_{conservation} \qquad [3]$$

In general such changes might also indicate changes in the regional ecosystem's functioning. It can be concluded that the assignment of both values to two individual scales is somewhat fuzzy. This is because there is a continuous transition between the scale of assessment and the assessed type of value. However, the motivation for taking local species richness into account is more its value for ecosystem functioning and the motivation for taking regional species richness into account is more its conservation value.

Generally speaking, different combinations of local species richness and the degree of regional threat are possible (Tab. 7-1). Total values for the combinations also vary. However, I could determine neither the degree of rarity nor the threat posed to species. This is a shortcoming of the proposed method. Instead, I distinguished between the change in the richness of all the species and the change in richness of threatened species only.

Tab. 7-1: Possible combinations of local richness and regional degree of threat.

"Scenario"	Local scale	Regional scale	Value	Example
Best case	many species	species found are threatened	+++	low-intensity meadows
Ecosystem functioning	many species	species found locally are common in the region	++	edges of mixed forests, hedgerows
Red list	only few species	species found are threatened	++	peat bogs
Worst case	only few species	species found locally are common in the region	+	industrial plants, arable fields

7.3 *Classification of land use types*

In order to assess land use activities, a classification of land use types[7] has to be set up. I used the CORINE land cover classification (European Environmental Agency 2000, cf., Appendix 1). The CORINE typology includes all the major land cover types of Europe and provides three different levels of classification (Tab. 7-2). Consistent maps are planned all of Europe.

Some modifications were necessary because the classifications in their original form did not distinguish between low- and high-intensity land use. In cases like forestry and agriculture, however, it is very important to distinguish high-intensity from low-intensity land use. For LCA purposes, it is obviously important to have different characterisation factors for, for example coniferous plantations and forests, which are close-to-nature. The modified classification CORINE+ is given in Appendix 1.

Switzerland uses another classification for its land use statistics. To compare the CORINE+ and the Swiss land use statistics, I set up a third classification with only 8 intensity types of land use. For this the Hegg Atlas of Threatened Vegetation Types (Hegg et al. 1993) was used (chapter 7.6.2).

The assignment of specific types of grassland to the CORINE+ types *pastures and meadows* (ID 231) and *semi-natural grassland* (ID 321) was performed according to Ellenberg (1996, pp. 786). The criterion for the assignment was the extent of fertilisation. Arrhenatheratalia (oat grass meadow), Calthion (fertilised moist meadow), and Polygono (oat grass meadow) are fertilised and were classified as *pastures and meadows*. Violion (mat-grass meadow), Xero-Bromium (xeric grassland), Meso-Bromium (mesoxeric grassland), Filipendula

[7] The difference between a land use type and a land cover type is that the former refers to the ongoing activities on a piece of land and the latter refers to the result of those activities. Despite this difference the two are used as synonyms in many publications including my own.

(tall-herb communities), and Molinion (moor-grass meadow) are generally not fertilised and were assigned to the category *semi-natural grassland*. When types of vegetation of the *semi-natural grassland* category are fertilised, they change – in dependency on moisture and acidity of soil – to one of the types in the *pastures and meadows* group.

Tab. 7-2: Coarse classification of the CORINE land use classification (detailed classification in appendix 1)

1 Artificial surfaces	**3 Forests and semi-natural areas**
11 Urban fabric	31 Forests
12 Industrial, commercial, and transport	32 Shrub and/or herbaceous vegetation associations
13 Mine, dump, and construction sites	33 Open spaces with little or no vegetation
14 Artificial, non-agricultural areas with vegetation	
2 Agricultural areas	**4 Wetlands**
21 Arable land	41 Inland wetlands
22 Permanent crops	42 Coastal wetlands
23 Pastures	
24 Heterogeneous agricultural areas	**5 Water bodies**
	51 Inland waters
	52 Marine waters

Another classification of land use types is used in the Ecoinvent (Frischknecht et al. 1996, chapter 8 pp. 20). It makes a distinction between 4 ecosystem types according to the hemerobian scale:

- Ecoinvent type I: natural land, impact of human beings not larger than those of other species
- Ecoinvent type II: modified land, impact of human beings larger than those of other species, but has non-cultivated components (e.g., near-to-nature forests)
- Ecoinvent type III: cultivated land, impact of human beings larger than those of other species, mostly cultivated components (e.g., agriculture, forest plantations)

- Ecoinvent type IV: built up land, dominated by streets, buildings, industrial plants.

This classification is used in the case study, since the inventory is derived from the Ecoinvent.

7.4 *Land occupation/transformation and assignment of damages*

The basic land use activities are occupation and transformation

In Life Cycle Assessment, land occupation and land transformation can be distinguished as basic land use activities (chapter 6.1, pp. 41). They result in either damage to or benefits for ecosystem quality (EQ). Whereas, land transformation causes a change in EQ, land occupation postpones changes to EQ. The restoration of land cover types, like peatbogs, is seen as a special type of transformation (long transformation times, with or without human activity).

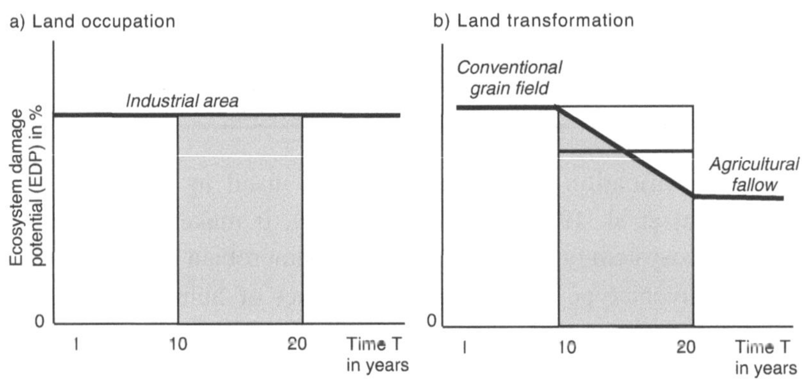

Fig. 7-2: Basic land use activities in Life Cycle Assessment. The damage (grey area in the graph) equals the level of damage potential multiplied by the time of occupation or transformation. Please note that in graph b land restoration, which is a particular type of transformation, is shown. The horizontal line (starting at a value of 10

years and ending at 20 years) refers to the mean EDP calculated in equation [5].

The damaging potential of land occupation refers to the use of a piece of land which shows a constant quality over time. This potential increases as the ecological quality of the occupied land decreases (Fig. 7-2a). In the case of land transformation, two aspects should be distinguished: (i) changes of potential damages (improvement/ deterioration) in time as well as (ii) potential damages integrated over time.[8] In the example of Fig. 7-2b, the potential damage decreases in time (i.e., ecosystem quality of agricultural fallow is higher compared to that of conventional grain fields). At the same time the potential damage integrated over time (equals the grey area in the graph) is rather large, since the individual plot is occupied by a series of land use types continuously changing from bad (conventional grain field) to moderate (agricultural fallow).

For land occupation the potential damage integrated over time taking species diversity into account (D_{occ}^{sp-div}) can be calculated as

$$D_{occ}^{sp-div} = EDP_{occ}^{sp-div} * Area * Time_{occ} \qquad [4]$$

where the ecosystem damage potential (EDP_{occ}^{sp-div}) is the characterisation factor and $Time_{occ}$ is the duration of occupation.

The potential damages integrated over time for a transformation (D_{trans}^{sp-div}) equals

$$D_{trans}^{sp-div} = \left(\frac{EDP_{ini}^{sp-div} + EDP_{fin}^{sp-div}}{2} \right) * Area * Time_{trans} \qquad [5]$$

8 In some publications a change of the damage potential in time is named transformation impact and the damage integrated over time is named occupation impact (e.g., Lindeijer et al. 2001). In their terminology an occupation causes an occupation impact. A transformation, however, triggers both a transformation impact and an occupation impact.

where EDP_{ini}^{sp-div} is the initial ecosystem damage potential and EDP_{fin}^{sp-div} is the final damage potential and $Time_{trans}$ is the duration of transformation.

The delta change of potential damages ($\Delta D_{trans}^{sp-div}$) between to points in time ($T_{initial}$ and T_{final}) resulting from a transformation can be calculated as

$$\Delta D_{trans}^{sp-div} = \left(-EDP_{ini}^{sp-div} + EDP_{fin}^{sp-div}\right) * Area \qquad [6]$$

where positive results refer to deterioration and negative results to improvements in terms of ecosystem quality. The same applies for potential benefits which occur when occupied land or final land after transformation shows high ecosystem quality. For the sake of simplicity, I will replace EDP^{sp-div} with EDP and D^{sp-div} with D in all following chapters.

Sequence of land occupation and transformation

The equations [4], [5], and [6] are the basis upon which to calculate the damage/benefits of a more complex sequence of land occupation and transformation in a modular mode.

In the case of land use activities with a defined end of use phase and a restoration of the area used (e.g., mining), the total potenial damage (or benefits) should be calculated as the sum of the potenial damage of the transformation at the beginning, of the occupation phase in the middle and of the transformation (restoration) at the end (Fig. 7-3a). The potenial damage integrated over time from occupation and transformation are calculated according to the basic equations [4] and [5]. The delta change of the potential damage in time is zero in this example since initial EDP and final EDP are equal (equation [6]).

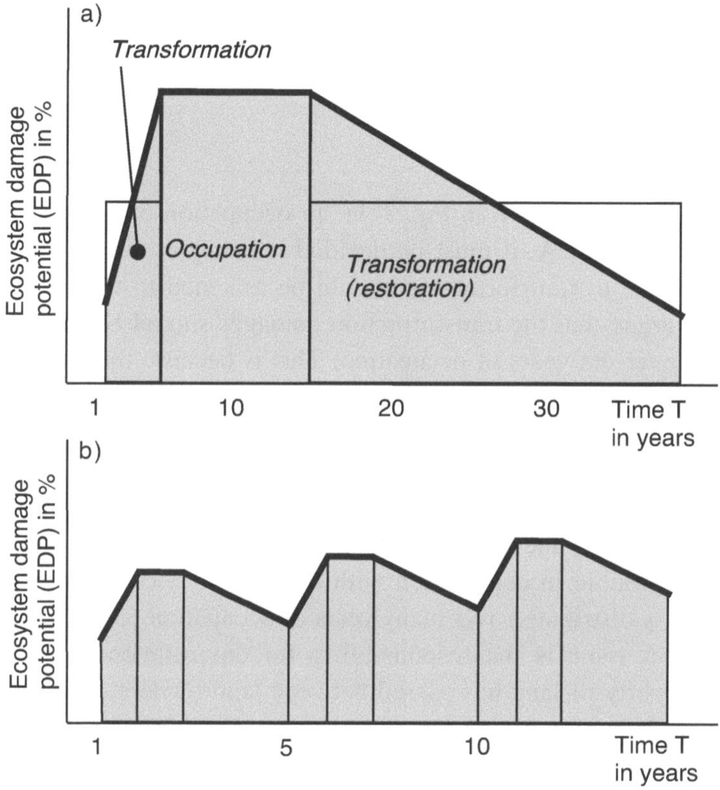

Fig. 7-3: Land use activities as sequences of occupation and transformation. Mining (a) and the rotation systems in agriculture (b) are two examples of such complex sequences of land use. The squares in graphic a) are the graphical representation of equation [5].

The ecosystem damage potential of a complex sequence of land occupation and transformation (e.g., rotation systems in agriculture, where periods of intensive use of a field and those of extensive use for recovery change periodically) can be calculated by adding up the individual potential damage resulting from occupation and transformation (Fig. 7-3b). In the example of Fig. 7-3b the delta change of damage potential in time for the total sequence is above zero since ecosystem quality gradually deteriorates and therefore, final EDP is worse than initial EDP.

Assignment of damages

If, as in the case shown in Fig. 7-3a, an occupation of 1 year were to be assessed in LCA, it must be decided whether or not a part of the damages due to transformation should be assigned to the occupation phase. I argue that the transformation damages should be equally distributed over the years of occupation. This is because the damage due to transformation would not occur without the occupation phase. Taking a mining case for example, it is quite obvious that all of the damage (not only that which is due to occupation) should be attributed to the amount of ore produced. If the occupation phase is very long (e.g., 500 years) the quantity of damage due to transformation becomes negligible in comparison with that due to occupation, because it is equally distributed over many years of occupation.

Another issue is the responsibility for environmental problems, when property of land has passed between land owners, e.g., after the first transformation phase. One could argue that subsequent landowners are not responsible for damage which their predecessor initiated. It could also be argued that the land transformer is not the only person responsible for the drop in EQ, but that subsequent land users should also carry a part of that damage. However, I think that the link between ownership and responsibility cannot be addressed in LCA. Being responsible for environmental damage does not require the ownership of a piece of land. For example, the people who demand and consume the good produced on that land are at least partially responsible. Therefore, the system borders should be drawn disregarding ownership. Tracing back the ownership of land requires site-specific information, which anyway might be hard to get in LCA.

Reversibility of damages

It can be argued that the land cover type should not necessarily be restored to the same type of land cover after occupation, but rather to one of similar value to the original one. If this is not possible (i.e., the loss in ecosystem quality is not reversible), then the damage is infinite.

An example of infinite damage is the clearcutting of tropical rain forest to raise cattle. The damage is infinite, because most likely ecosystem quality can never arrive at an equal level, and therefore, restoration time is infinite. One problem with this line of reasoning, however, is that this absolutely does not mean that clearcutting 1 ha of a rainforest is equal to clearcutting 100 ha. One practical solution for LCA could be to assume very long restoration times[9] for such a transformation and include those into LCA. Otherwise, one has to conclude that such damages cannot be adequately taken into account in LCA and corresponding land use activities have to be ceased, for example, by law.

7.5 *Local characterisation factor: ecosystem damage potentials for species diversity (EDP_{local})*

In this chapter the local characterisation factor for ecosystem damage potentials (EDP_{local}) with focus on species diversity is developed. In the effect analysis the species richness for a variety of land use types was determined. The applied method for this is described in chapter 7.5.1 (pp. 72) and the results are given in chapter 7.5.2 (pp. 83). In the damage/benefits analysis the observable effects are transformed into potential damage/benefits. The proposed method for calculating the

9 An example is the use of peatbogs to produce peat for gardening and heating. This land cover type became very rare in Western Europe and restoration time is between 1'000 and 10'000 years.

EDP_{local} can be found in chapter 7.5.3 (pp. 91). The numerical results for EDP_{local} are given in chapter 7.5.4 (pp. 93).

7.5.1 Local species richness of specific land use types

The species-area relationship

To derive reliable characterisation factors, it is necessary to explicitly consider the nonlinearity of the relationship between area and species number. Arrhenius (1921) was one of the first scientists to describe the "...obvious fact, that the larger the area taken, the greater the number of species". The species-area relationship is observable in both the size of homogeneous land cover types and the size of whole regions which consist of a mosaic of different land cover types. Generally, two hypotheses are put forward to explain the correlation between the species number and the size of the area (i.e., patch size, but also size of a region) – the habitat-diversity hypothesis and the area-per-se hypothesis (Preston 1962; Connor and McCoy 1979).

According to the habitat-diversity hypothesis, the connection between increases in the size of an area and the species number is explained by the fact that when investigating successively larger areas, the number of habitats in the sample increases.

In contrast, the area-per-se hypothesis emphasises immigration and extinction rates. Assuming that large areas contain large numbers of members of a species and that local species extinction is dependent on the number of members (i.e., the higher the number of species members, the lower their extinction probability), then in large areas more species should be observed than in small areas. Moreover, an observed area's or patch's degree of isolation is important for this hypothesis (for more details on both hypotheses see Preston 1962; Connor and McCoy 1979).

It is likely that both mechanisms determine the dependency of the species number on the area size; however, the relative importance of both hypotheses depends on the specific case. For "continental" species-area relationships, the habitat-diversity hypothesis is probably more important than for "true island" species-area relationships.

Arrhenius (1921) derived a function for the species-area relationship (SPAR):

$$S = aA^b \qquad [7]$$

$$\ln S = \ln a + b \ln A \qquad [8]$$

where S is the species number, A the area of the plot, and a (measure for species richness) and b (measure for species accumulation rate) are the parameters of the function.

The log-transformed power model (equation [8]) is assumed to fit the empirical data sets on species numbers quite well as a function of area. However, other functions ($S = aA + b$; $\ln S = aA + b$; $S = a + b \ln A$) for fitting SPARs have also been proposed (Connor and McCoy 1979). Connor and McCoy (1979) conducted a study in which the quality of the fit of these four possible models and 100 empirical data sets was tested. Although they concluded that for each data set the appropriate model has to be chosen individually, the power model ($\ln S = \ln a + b \ln A$) resulted in 75 cases with "... no substantial lack-of-fit ..." (Connor and McCoy 1979, p. 795), of which the power model fits best in 36. Based on this study, I conclude that the log-transformed power model is an appropriate model for the species-area relationship.

The transformed power model has two parameters. The parameter $\ln a$ (y-intercept) indicates the species richness of a sample standardised for $A=1$. The parameter b (slope) denotes species accumulation rates. The species richness (parameter $\ln a$) is supposed to be dependent on ecological diversity within a land cover type or region. The species accumulation rate (parameter b) for a region depends inter alia on the degree of isolation. For true islands with a high degree of isolation, the slope-parameter is in the range of 0.20-0.40 (Preston 1962), while for continental patches with incomplete isolation, the slope is found to be between 0.12-0.19 (Connor and McCoy 1979).

Meta-analysis of species richness

My objective was to make different land use types comparable with respect to their species richness by applying a meta-analysis. As a basis two types of empirical data exist. First, species numbers on specific land use types are determined for plots with observable borders in the landscape (e.g., industrial areas within rural land) which vary in size (*type I* data). Second, species are counted on standardised plots of equal size, e.g., 100 m^2 (*type II* data), cf., Fig. 7-4a. Generally speaking, two options exist for the meta-analysis:

- **Option 1:** Individual species-area relationships (SPAR)
  The first possibility is to construct one SPAR for each land use type. For the *type I* data, it is possible to directly fit SPARs because of the variation in plot sizes (equation [7]). In the second case (*type II* data), there is no variation in plot sizes. In this case, the rarefaction method (Hulbert 1971; Heck et al. 1975), could be used. The expected numbers of species $E(S_n)$ is calculated by randomly choosing a sub-sample n from all N plots in the sample:

$$E(S_n) = S - \frac{\sum_{i=1}^{S} \binom{N - N_i}{n}}{\binom{N}{n}} \qquad [9]$$

  where N is the total number of plots in the sample, N_i the number of plots where species i is found, n the number of randomly chosen plots, and S the total number of species on all the plots. Given a fixed number of species S and a fixed number of plots n, the expected number of species is dependent on the species' abundance (the more abundant a species is, the more plots N_i are inhabited by it). The expected number for *n* plots increases with the increase in abundant species, since it is more likely that a common species is found than a rare species.

A power function could then be fitted to the discrete rarefaction function in order to determine the parameters of the continuous function. Therefore, instead of calculating the species numbers for defined areas, the parameters of equation [7] could be used to compare species richness of different land use types.[10]

However, option 1 has a major disadvantage. It is necessary to calculate SPARs on very different scales. The local scale is 100 m^2 to 1 km^2 and the regional scale is 100 km^2 to 10'000 km^2. The *type I* data available on the specific land use types differs quite substantially and is mostly available for a narrow spatial scale. To account for this, individual extrapolation for each land use type would be necessary in option 1, and doing so would increase uncertainty to such an extent that this is not appropriate. Therefore, I developed option 2 as follows.

- **Option 2:** Standardisation of the species richness with a single SPAR

  The second option is to standardise the species number S using one single species-area relationship for all land use types. To do this I adjusted the species-area relationship according to equation [8], which yields a straight regression line on a ln-ln scale, that fits all of the empirical data together. Next I calculated the standardised species number $S_{standard}$ by shifting the data points, in parallel with the regression line, to the standard area size (Fig. 7-4a). This eliminates the aspect of *S* which is attributable to area

[10] Lindeijer et al. (1998, pp. 15 and Annex 1.7) propose a species accumulation factor for assessing local land use impacts on biodiversity. Their accumulation factor represents the increase in species number, which occurs when the area in-

size. In addition this procedure has the advantage that all data points are treated equally.

The standardised species number ($S_{standard}$) is calculated as

$$S_{standard} = S - \Delta S \text{ (see Fig. 7-4)} \quad [10]$$

where S is the original species number and ΔS is that part of the species number which can be attributed to the area rather than to the land use types.

Here ΔS is calculated as

$$\Delta S = S' - S'_{standard} \text{ (see Fig. 7-4)} \quad [11]$$

where S' is the average species number for the area A and $S'_{standard}$ is the average species number for the standard area $A_{standard}$. The average species number takes all land use types into consideration and is calculated using the regression line (Fig. 7-4b).

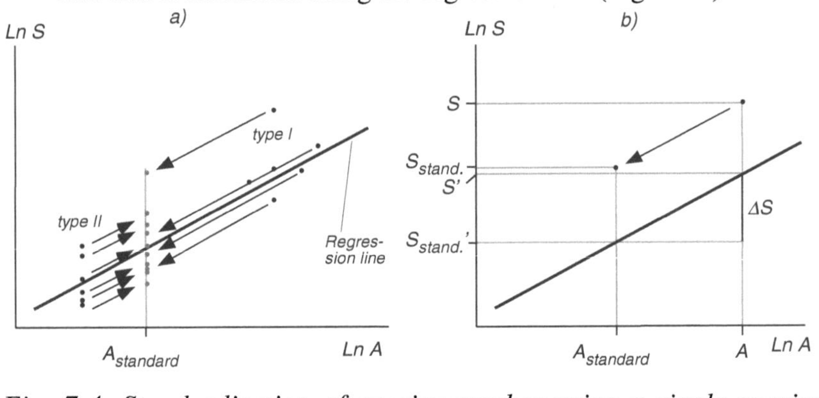

Fig. 7-4: Standardisation of species number using a single species-area relationship (regression line). a) The original data points for the species number are shown in black, the standardised in grey. (A is area; S is species) b) The calculation of the standardised species number ($S_{stand.}$) is shown for one data point (S' is average species number for area A).

---

creases by a factor of 10. This accumulation factor is taken as a measure for species richness.

Testing the independence of standard error of species number (ln S*) and the size of plots (ln* A*)*

The standardisation changes the spread of the data, because the parallel shift is done on a ln-ln scale. This could result in a bias of calculated uncertainties. Therefore, I tested whether the variance of the data is independent of the size of the plots. To do this, I have grouped all data on area ($\ln A$) and species number ($\ln S$) into classes of width 1 $\ln A$. I calculated the mean of $\ln S$, its standard deviation σ, and its standard error SE for each class of $\ln A$ separately. Then I determined the regression line for the $\ln A$ classes (x-axis) and SE (y-axis), see Fig. 7-7, as well as the 95% confidence interval for the parameter. I took the standard error

$$SE = \frac{\sigma}{\sqrt{n}}$$

where σ is the standard deviation and n is the sample size. For the test SE is more appropriate than σ, because σ varies with sample size n, which can differ considerably from class to class. In contrast, SE is corrected for the sample size. The results of this test are shown in chapter 7.5.2, p. 83.

Data for local biodiversity

In order to include data on the plant species richness of the individual land use types I performed a meta-analysis of published investigations. Most of the publications stem from vegetation science. Plant species and composition of the vegetation types were investigated according to the methods from Braun-Blanquet. The sources of empirical data which I took into consideration are given in Tab. 7-3. I provided a description of land use types which were investigated in the original publication, because those do not match with the CORINE+ classification exactly. The distribution of samples of land use types studied by the authors are given in the Tab. 7-4.

Tab. 7-3: Description of the types of land use for which empirical data was available and their relation to the CORINE+ land cover classification (cf., Appendix 1).

ID	Land use type	Data source	Description	Country
111	Continuous urban	(Reidl 1989)	City centre (almost totally sealed)	Germany (Essen)
		(Schulte 1985)	City centre (almost totally sealed), edge of the city centre	Germany (Bochum)
112	Discontinuous urban	(Reidl 1989)	Business and residential district with green areas and gardens.	Germany (Essen)
		(Schulte 1985)	Residential district with garden	Germany (Bochum)
113	Urban fallow	(Reidl 1989)	Remains of demolished houses are removed and areas are levelled.	Germany (Essen)
		(Sukopp 1990)	"	Germany (Berlin)
114	Rural settlement	(Schulte 1985)	Village with buildings, streets, and gardens	Germany (Bochum)
1212	Industrial area part with vegetation	(Reidl 1989)	High or medium proportion of sealed area; open soil, sometimes contaminated with chemicals.	Germany (Essen)
		(Schulte 1985)	"	Germany (Bochum)
1224	Rail embankments and associated land	(Reidl 1989)	Goods depots, stations, and railway embankments.	Germany (Essen)
		(Schulte 1985)	"	Germany (Bochum)
		(Sukopp 1990)	"	Germany (Berlin)
1225	Rail fallow	(Schulte 1985)	Fallow land with spontaneous vegetation.	Germany (Bochum)
125	Industrial fallow	(Reidl 1989)	Areas with remains of industrial buildings; deposits of rubble, gravel, sand	Germany (Essen)

ID	Land use type	Data source	Description	Country
			and industrial waste.	
134	Mining fallow	(Reidl 1989)	Levelled deposits of slag heaps (due to hard coal mining), rubble, gravel, and sand.	Germany (Essen)
141	Green urban	(Reidl 1989)	Parks, green areas, and cemeteries.	Germany (Essen)
		(Schulte 1985)	Parks and green areas	Germany (Bochum)
142	Sport and leisure facilities	(Albracht 1997)	Soccer pitch	Germany
2111	Intensive arable	(Schreiber 1995)	Area sown with cereals, conventional management.	Switzerland (Plateau)
		(Callauch 1981)	Root crops and cereals, conventional management.	Germany (Göttingen)
		(Bigler et al. 1998)	not specified	Switzerland (Plateau)
2112	Integrated arable	(Schreiber 1995)	Field for cereals, integrated management, i.e., reduced input of pesticides and fertiliser.	Switzerland (Plateau)
		(Bigler et al. 1998)	not specified	Switzerland (Plateau)
2113	Organic arable	(Schreiber 1995)	Area sown to cereals, organic management.	Switzerland (Plateau)
		(Callauch 1981)	Root crops and cereals, organic management.	Germany (Göttingen)
		(Bigler et al. 1998)	not specified	Switzerland (Plateau)
2114	Fibre/ energy crops	(Lips et al. 1997)	21141 Kenaf (Hibiscus cannabinus L.), 21142 Hemp (Cannabis sativa L.), 21143 Chinese reed (Miscanthus sp.),	Switzerland (Plateau)
2115	Agricultural fallow	(Bigler et al. 1998)	not specified	Switzerland (Plateau)
		Schwab	not specified	Switzerland (Plateau)

ID	Land use type	Data source	Description	Country
2222	Organic orchards	(Bigler et al. 1998)	not specified	Switzerland (Plateau)
2311	Intensive meadow	(Lips et al. 1997)	Meadows harvested mechanically 3 times per year, fertiliser applied, some on former arable land.	Switzerland (Plateau)
		(Bigler et al. 1998)	not specified	Switzerland (Plateau)
2312	Less intensive meadow	(Lips et al. 1997)	Meadows harvested mechanically 2 or 3 times per year, reduced input of fertiliser, some on former arable land.	Switzerland (Plateau)
		(Wittwer et al. 1997)	"	"
		(Bigler et al. 1998)	not specified	Switzerland (Plateau)
2313	Organic meadow	(Lips et al. 1997)	Meadows harvested mechanically 1 time per year, no fertiliser, some on former arable land	Switzerland (Plateau)
		(Wittwer et al. 1997)	"	"
		(Bruelheide 1995)	Meadows harvested mechanically 1 time per year; Arrhenatheretalia (Oat grass meadow), Calthion (Fertilised moist meadow), Polygono (Oat grass meadows)	Germany (Harz)
		(Grüttner 1990)	Calthion (Fertilised moist meadow),	Germany (Bodensee)
		(Bigler et al. 1998)	not specified	Switzerland (Plateau)
245	Agricultural fallow with hedgerows	(Bigler et al. 1998)	not specified	Switzerland (Plateau)
311	Broad-leafed forest	(Wohlgemuth 1992)	Broad-leafed forest with mainly beech (Fagus sylvatica over 50%).	Switzerland (Plateau)

ID	Land use type	Data source	Description	Country
		(Döring-Mederake 1991)	Moist forests: Carici (Fen alder wood), Pruno-Fraxinetum (Alder ash wood), Querco-Ulmetum (Oak elm wood), Rubus idaeus-Alnus glutinosa (Alder wood), Vaccinium uliginosum-Betula pubescens (Fen birch wood)	Germany (northwest)
314	Forest edge	(Flückiger 1999)	Edges of mixed broad-leafed, coniferous and broad-leafed forests.	Switzerland (Plateau)
		(Bigler et al. 1998)	"	"
321	Natural grassland	(Manz 1997)	Violion (Mat-grass pasture), Xero-Bromium (Xeric grassland), Meso-Bromium (Mesoxeric grassland)	Germany (Rheinhessen, Pfälzer Wald)
		(Bruelheide 1995)	Violion (Mat-grass pasture)	Germany (Harz)
		(Grüttner 1990)	Filipendula (Tall-herb communities), Meso-Bromium (Mesoxeric grassland), Molinion (Moor-grass meadow)	Germany (Bodensee)
322	Heathland	(Manz 1997)	Dwarf-shrub heaths	Germany (Rheinhessen, Pfälzer Wald)
325	Hedgerows	(Bigler et al. 1998)	not specified	Switzerland (Plateau)
412	Peatbogs	(Bruelheide 1995)	Caricion (small-sedge meadow)	Germany (Harz)
		(Grüttner 1990)	Peatbogs, sedge swamps, fen woods	Germany (Bodensee)
	Swiss Plateau	(Wohlgemuth 1993; WSL/FNP without year)	Northeastern part of Swiss Plateau with current mix of land uses (about 20% of total area is low-intensity land use).	Switzerland (Plateau)

Tab. 7-4: Number of sample plots per source and land use type.

	Albracht	Bruelheide	Callauch	Döring	Flückiger	Grüttner	Bielger et al.	Lips et. al.	Manz	Reidl	Schreiber	Schulte	Schwab	Sukopp	Wittwer	Wohlgemuth	Totals
Agricultural fallow							18						120				138
Agricultural fallow with hedgerows							11										11
Continuous urban										10		50					60
Conventional arable			38								16						54
Discontinuous urban										59		61					120
Fibre/energy crops (Chinese reed)								50									50
Fibre/energy crops (hemp)								20									20
Fibre/energy crops (kenaf)								24									24
Forest edge					27		51										78
Green urban										75		35					110
Heathland									6								6
Hedgerows							44										44
Industrial area part with vegetation										29		24					53
Industrial fallow										26							26
Integrated arable							85	54			18						157
Intensive meadow							51	18									69
Less intensive meadow							51	20							33		104
Mining fallow										10							10
Natural grass-land		78				120		4	40								242
Organic arable			50								12						62
Organic meadow		462				100	112	25							28		727

	Albracht	Bruelheide	Callauch	Döring	Flückiger	Grüttner	Bielger et al.	Lips et. al.	Manz	Reidl	Schreiber	Schulte	Schwab	Sukopp	Wittwer	Wohlgemuth	Totals
Organic orchard							48										48
Peatbog		28				605											633
Rail embankments										41		13		3			57
Rail fallow												19					19
Rural settlement												17					17
Semi-natural broad-leafed forest (arid)																223	223
Semi-natural broad-leafed forest (moist)				501													501
Sport facilities	16																16
Urban fallow										24				3			27
Total	16	568	88	501	27	825	471	215	46	274	46	219	120	6	61	223	3706

7.5.2 Results for local effects

The species-area relationship calculated for the standardisation

To standardise the species numbers, I calculated the regression line for all local plots (homogenous land use types) and for the regional plots (mix of land use types). The function is $\ln S = 4.1 + 0.2 \ln A$ in ha and has a correlation coefficient $R^2 = 0.64$. Size and species number of all local plots and regional polygons of Switzerland (Fig. 7-10, p. 98) investigated are shown in Fig. 7-5. Based on this graph the standardised species number was calculated as proposed in chapter 7.5.1 (pp. 72).

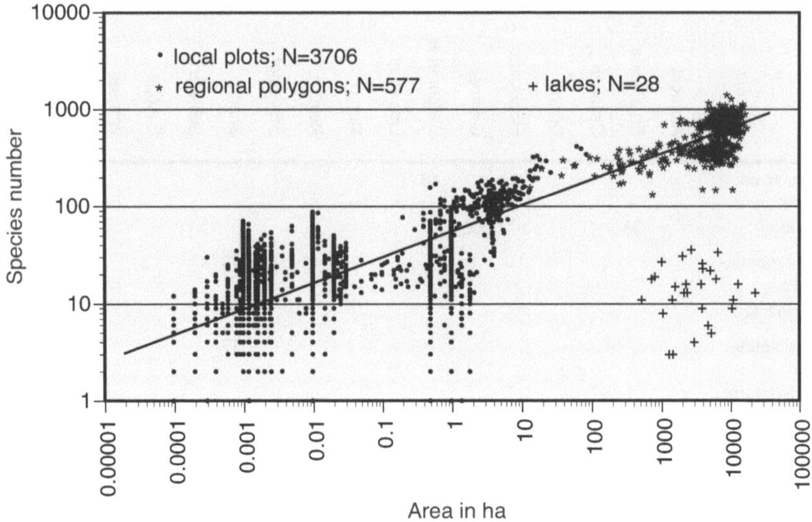

Fig. 7-5: Regression (lnS=4.1+0.2lnA, R^2=0.64) for the area in ha and species number taking local plots and regional polygons into account.

A refinement of the standardisation based on the total regression would be to calculate individual regressions for high-intensity land use, low-intensity land use, as well as non-use and to use those results for standardisation. Species accumulation rate of high-intensity land use is expected to be lower compared to low-intensity land use or non-use. This is because high-intensity land use types are assumed to be more homogenous in terms of habitat diversity compared to low-intensity land use or non-use. According to the habitat-diversity hypothesis species accumulation rate is dependent on size of the area sampled because large areas show more habitats, which in turn inhabit more species.

The regression lines were calculated as follows:

for high-intensity land use:

$\ln S = 3.3 + 0.21 * \ln A$ in ha; $R^2 = 0.26$,

for low-intensity land use:
$\ln S = 3.8 + 0.09 * \ln A$ in ha; $R^2 = 0.18$, and

for non-use:
$\ln S = 4.5 + 0.30 * \ln A$ in ha; $R^2 = 0.45$.

Correlation coefficients were much lower compared to this of the total regression. In contrast to expectation low-intensity land use had a much lower accumulation rate (equals the steepness of the curve, which is 0.09 in this case) than high-intensity land use (0.21). Because these results are rather uncertain and counterintuitive, I decided to rely on the total regression to calculate the standardised species number. However, a possible explanation for this might be that average plot sizes differ.

Results of the standardisation

I have calculated the standardised species number for 0.01 ha ($A_{stand.}$), because the mean of ln*A* for all plots is -4.45 (equal to 0.0167 ha). The unstandardised species numbers and the standardised ones are shown for specific land use types, for Swiss regions, for types of intensity of land use, and for Ecoinvent types (Tab. 7-5). The exact assignment of land use types to the 8 types of intensity as well as to the 3 Ecoinvent types are given in appendix 4.

The standardisation results in a ranking of land use types and regions which is independent of the area I have chosen for the standardisation. Though the absolute species number changes with the standardisation size, the relative species number $S_{land\ use\ type}/S_{region}$ remains constant.

In general the results meet the expectations. However, the intensity types *high-intensity artificial area* (artificial_hi) and *low-intensity forests* (forest_li) show surprisingly almost equal standardised species numbers. According to appendix 4 more specific land use types within these types are for example discontinuous urban land (artificial_hi) and semi-natural broad-leafed forest (forest_li). Both specific land use types also show very similar species numbers, which are 22 and 23, respectively.

Tab. 7-5: Unstandardised and standardised species numbers S for specific land use types, Swiss regions, 8 types of intensity of land use, and Ecoinvent types (Frischknecht et al. 1996, chapter 8, pp. 20). The area chosen for standardisation ($A_{stand.}$) was 0.01 ha. Within each group types are ranked according to standardised species number.

		Unstandardised species number (*S*)							Standardised *S* for $A_{stand.}$ 0.01 ha					
CORINE+ No.	Type	Mean	Std. Dev.	Std. Error	Count	Min	Max	Median	Mean	Std. Dev.	Std. Error	Min	Max	Median
Specific land use types														
142	Sport facilities	17	7	1.7	16	8	35	17	7	3	0.7	3	14	7
2112	Integrated arable	10	8	0.7	157	1	39	7	7	5	0.4	0	26	5
111	Continuous urban	21	12	1.5	60	1	52	21	8	4	0.5	0	16	8
2111	Conventional arable	13	7	0.9	54	1	32	12	10	6	0.9	1	25	9
21141	Fibre/energy crops (kenaf)	22	11	2.3	24	1	44	23	10	5	1.0	0	20	10
21142	Fibre/energy crops (hemp)	24	9	2.1	20	9	45	23	11	4	0.9	4	20	10
21143	Fibre/energy crops (chinese reed)	33	17	2.5	50	1	65	34	15	8	1.1	0	29	15
2311	Intensive meadow	16	6	0.8	69	5	38	15	17	9	1.1	2	38	17
322	Heathland	13	3	1.4	6	9	18	13	18	4	1.8	12	24	17
2312	Less intensive meadow	24	7	0.7	104	9	39	24	19	11	1.1	4	46	16
412	Peatbog	12	7	0.3	633	1	38	11	19	10	0.4	2	62	17
112	Discontinuous urban	68	34	3.1	120	16	137	63	22	9	0.8	6	39	22
2212	Organic orchard	20	6	0.9	48	11	35	20	23	7	1.0	13	43	22
3112	Semi-natural broad-leafed forest (arid)	25	10	0.7	223	5	51	24	23	10	0.6	4	51	22
1212	Industrial area, part with vegetation	82	53	7.2	53	5	183	58	24	13	1.8	2	50	22
1225	Rail fallow	61	14	3.3	19	38	89	60	24	6	1.3	15	34	23

CORINE+ No.	Type	Unstandardised species number (S) Mean	Std. Dev.	Std. Error	Count	Min	Max	Median	Standardised S for $A_{stand.}$ 0.01 ha Mean	Std. Dev.	Std. Error	Min	Max	Median
3112	Semi-natural broad-leafed forest (moist)	24	10	0.4	501	4	70	22	24	10	0.4	4	70	22
114	Rural settlement	63	13	3.2	17	36	80	68	25	5	1.2	14	31	26
2113	Organic arable	28	7	0.9	62	6	39	29	26	10	1.2	2	39	29
141	Green urban	94	65	6.2	110	13	306	91	29	15	1.5	5	65	30
1224	Rail embankments	111	77	10.2	57	9	417	98	32	13	1.7	3	72	31
134	Mining fallow	150	31	9.8	10	108	197	139	38	6	1.9	28	49	37
321	Natural grassland	26	9	0.6	242	4	54	25	39	15	0.9	4	80	38
113	Urban fallow	104	52	10.1	27	48	325	92	40	13	2.6	21	74	36
125	Industrial fallow	144	60	11.7	26	72	344	136	40	11	2.2	20	57	42
2115	Agricultural fallow	38	17	1.5	138	8	86	37	43	14	1.2	12	83	43
325	Hedgerows	38	18	2.8	44	4	88	36	44	22	3.4	4	88	48
2313	Organic meadow	31	11	0.4	727	7	71	30	45	20	0.8	5	114	45
314	Forest edge	38	18	2.1	78	2	84	35	48	24	2.7	2	102	47
245	Agricultural fallow with hedgerows	53	6	1.8	11	45	65	50	53	6	1.8	45	65	50
Regions in Switzerland														
	Lakes	16	9	1.7	28	3	36	14	1	1	0.1	0	3	1
	Mountains above timberline	354	78	5.3	215	134	551	360	27	8	0.5	9	56	26
	Plateau	679	106	10.4	104	478	1078	659	40	7	0.7	28	62	39
	Jura	718	114	17.2	44	490	950	728	44	7	1.1	30	59	45
	Alps	809	131	9.2	202	517	1411	795	50	8	0.6	31	84	49
Types of intensity														
	agri_hi	18	13	0.6	436	1	65	15	13	9	0.5	0	39	11
	artificial_hi	57	42	2.7	249	1	183	39	18	11	0.7	0	50	15
	forest_li	24	10	0.4	724	4	70	23	23	10	0.4	4	70	22
	non-use	28	34	1.1	992	1	344	16	27	18	0.6	2	102	23
	artificial_li	96	67	5.0	184	9	417	87	29	14	1.0	3	72	30
	agri_li	29	10	0.3	1121	4	71	28	40	20	0.6	4	114	39

		Unstandardised species number (S)							Standardised S for $A_{stand.}$ 0.01 ha					
CORINE+ No.	Type	Mean	Std. Dev.	Std. Error	Count	Min	Max	Median	Mean	Std. Dev.	Std. Error	Min	Max	Median
Ecoinvent types														
	Type IV	57	42	2.7	249	1	183	39	18	11	0.7	0	50	15
	Type II	26	25	0.6	1958	1	344	20	27	16	0.4	2	102	24
	Type III (excluded)	96	67	5.0	184	9	417	87	29	14	1.0	3	72	30
	Type III	26	12	0.3	1315	1	71	25	32	22	0.6	0	114	28

Testing independence of standard error of ln S and size of plots ln A

A problem of the standardisation might be that the uncertainty range of data for the species number is heavily influenced by the standardisation (Fig. 7-6). Standardisation of a specific data set has no influence on the range of uncertainty on the logarithmic scale. At the same time, this means that data set's changes of the uncertainty on the non-logarithmic scale can be quite large. Take the data on organic meadows as an example (Fig. 7-6). It can be observed that before standardisation the uncertainty range (here shown as a boxplot with percentiles) is very narrow. After standardisation it is quite broad. This is only appropriate if the assumption that the range of uncertainty of the species number ln$S$ is independent of the size of the plots ln$A$ – where both are on a logarithmic scale – holds true. Therefore, I grouped all data of the local plots into classes of width 1 ln$A$ and calculated for each class the mean and uncertainty range (Tab. 7-6). More specifically, I took the standard error as a measure for the uncertainty range.

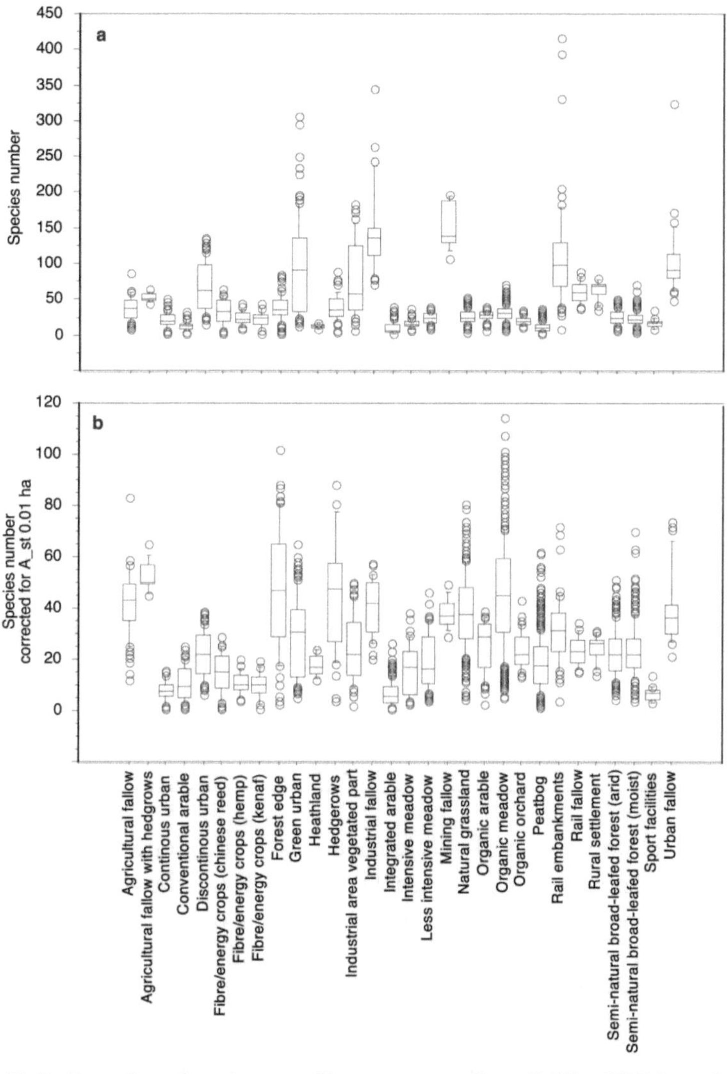

Fig. 7-6: Boxplot showing median, percentiles (25%, 75%) and outliers for unstandardised species numbers (graph a) as well as standardised ones (graph b).

Tab. 7-6: For each class of plot size (lnA) in ha information, such as the mean, standard deviation, and error is given for the unstandardised data of the species number lnS.

Count	class ln*A*	ln*S*				
		Mean	Std. Dev.	Std. Error	Min	Max
2	-12	0.69	0.00	0.00	0.69	0.69
3	-11	1.26	1.20	0.70	0.00	2.40
17	-10	1.62	0.59	0.14	0.00	2.48
30	-9	2.05	0.72	0.13	0.00	3.40
130	-8	2.23	0.60	0.05	0.69	3.56
1385	-7	3.01	0.69	0.02	0.00	4.26
98	-6	2.86	0.70	0.07	0.69	4.14
1039	-5	3.08	0.64	0.02	0.00	4.48
168	-4	3.10	0.42	0.03	1.61	3.91
15	-3	2.97	0.24	0.06	2.56	3.33
19	-2	3.16	0.44	0.10	2.71	4.34
277	-1	3.10	0.88	0.05	0.00	5.15
335	0	3.48	0.86	0.05	0.00	5.22
145	1	4.59	0.39	0.03	3.22	5.24
38	2	5.15	0.30	0.05	4.61	5.81
2	3	5.56	0.19	0.13	5.42	5.69
3	4	5.95	0.10	0.06	5.84	6.03

The null hypothesis is that standard errors are dependent from the class ln*A* which refers to the scale on which the data were sampled. The corresponding regression plot for classes of ln*A* and the standard error are given in Fig. 7-7. It shows that aside from one outlier from the class ln*A*=-11, the standard error for all other classes is quite similar. The 95% confidence interval for the slope coefficient of the regression line includes the zero value (lower border: -0.03, upper border: 0.01), which means that no dependency is given between class ln*A* and standard error. The conclusion of this test is that the uncertainty in species number is independent from the plot size and therefore the standardisation done is appropriate in terms of uncertainty ranges.

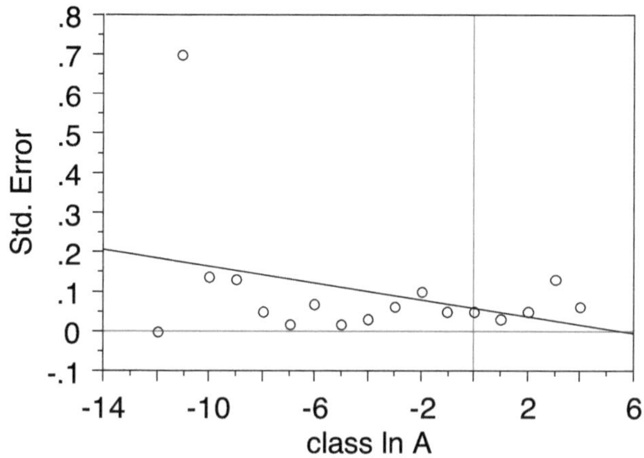

*Fig. 7-7: Regression line for the plot size lnA in ha (grouped into classes) as independent variable and standard error (Std. Error = 0.06 - 0.01 * class ln A; R^2 = 0.11) as dependent variable.*

7.5.3 Transformation of local effects into damage/benefits

The ecosystem damage potential on the local (EDP_{local}) scale is based on the results of the effect analysis. The observable effects which result from land use are the differences or changes in the species number per area. However, absolute species numbers are less meaningful than relative species numbers where a comparison with a reference is made. Different possibilities for the reference exist (chapter 6.4). I chose regional average species richness as my reference for assessing species richness on local plots. Species richness on a biogeographical scale varies remarkably. If one divides Europe into 4 diversity zones, the number of vascular plant species per 10'000 km^2 ranges from 200 to 500 in northern Scandinavia and from 2'000 to 3'000 in southern parts of Mediterranean countries (Barthlott 1998, p. 36). Obviously the impact of occupying a plot of land (with e.g., 10 species) should be assessed relative to the region where the occupation takes place.

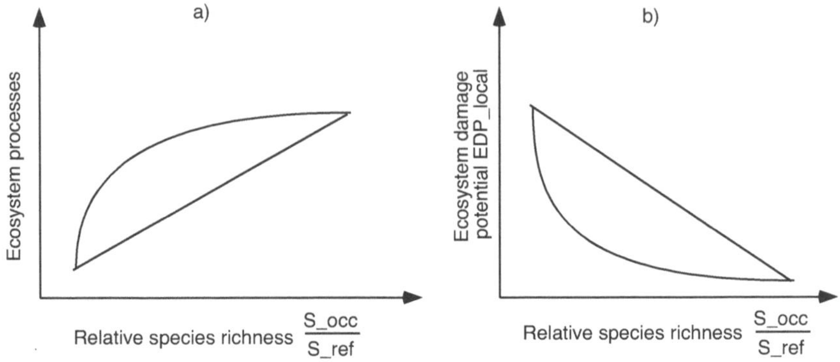

Fig. 7-8: a) Linear and logarithmic relationships between species richness and ecosystem processes. b) The corresponding curves for the effect-damage function on the local scale (S_{occ} is the species number on the occupied plot and S_{ref} that of the reference).

I considered two options for the effect-damage function: (1) a linear function and (2) a logarithmic one. Both functions are proposed to describe the functional relationship between species richness on a plot and ecosystem processes (Schläpfer and Schmid 1999) and are based on theories in ecosystem science (Schulze and Mooney 1994a). The logarithmic function (Fig. 7-8a) is supported by the redundant species hypothesis, i.e., that is, that the addition of one species results in a decrease in the marginal growth of utility in terms of ecosystem processes. Schläpfer et al. created an expert questionnaire and compiled the result with the finding that the logarithmic relationship is the most likely one (see also chapter 5.1.2, pp. 27, section *Correlation between ecosystem attributes*).

Based on these functions from ecosystem theory, I derived the following effect-damage functions. Thereby, I used the regional species richness of the Swiss regions as a reference.

Option 1: Linear effect-damage function

The ecosystem damage potential EDP_{local} can be calculated using the linear relationship as shown in Fig. 7-8b:

$$EDP_{local} = 1 - \frac{S_{occ}}{S_{region}} \quad [12]$$

where S_{occ} is the species number of an occupied land use type, S_{region} is the average species number in the region.

Option 2: Nonlinear effect-damage function

The other function for EDP_{local} is (also Fig. 7-8b)

$$EDP_{local} = 1 - \left\{ a \ln\left(\frac{S_{occ}}{S_{region}} \right) + b \right\} \quad [13]$$

where a and b are parameters. Although this function has the disadvantage that parameters a and b have to be assessed, this was possible using the work done by Schläpfer et al. (1999).

7.5.4 Results for EDP_{local}

The basis for calculating the local damage is the species number of a specific land use type S_{occ} standardised for 0.01 ha. This absolute species number is transformed into a relative number using the regional species richness as a reference (Tab. 7-7). The Swiss Plateau and the Central Alps were chosen as exemplary regions, because they are very different from one another in terms of land use intensity.

Linear and logarithmic transformations of the relative species numbers result in ecosystem damage potentials (*EDP*) for the exemplary regions (Tab. 7-7). The parameters needed in the logarithmic function (equation [13]) were assessed based on the work of Schläpfer et al. (1999). The parameters of the function was estimated as

$$\textit{Ecosystem processes} = 0.27 \ln (S_{occ}/S_{ref}) + 1 \quad [14]$$

and subsequently is the EDP_{local} according to equation [13]

$$EDP_{local} = 1 - (0.27 \ln (S_{occ}/S_{ref}) + 1) = - 0.27 \ln (S_{occ}/S_{ref}) \qquad [15]$$

Using these parameters, the curve (Fig. 7-9a) approximately has the shape estimated by the 39 experts in Schläpfer et al. (1999). The estimate of the experts was the following (the exact numbers calculated with equation [14] are given in brackets):

- 10% of the species sustain 40% (38%) of ecosystem processes
- 25% " 60% (63%) "
- 50% " 80% (81%) "
- 100% " 100% "

In this expert assessment differences in the importance of species for ecosystem processes were not considered by the experts. Therefore, this assessment has a rather course and preliminary status.

The resulting effect-damage curve I used to transform the relative species number is shown in Fig. 7-9b.

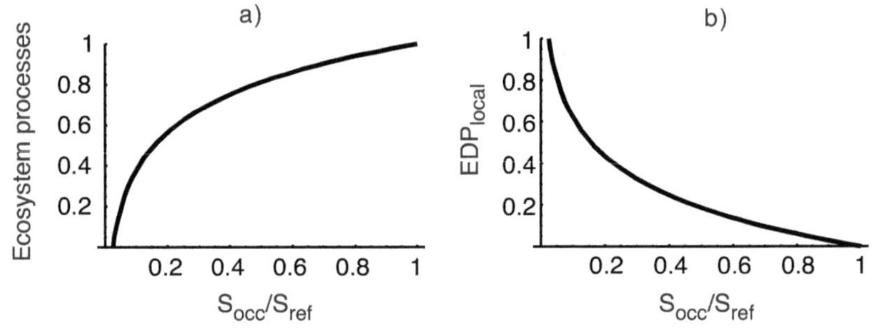

Fig. 7-9: a) Relationship between relative species number (S_{occ}/S_{ref}) and ecosystem processes as assessed in Schläpfer et al. (1999) (cf., [14]). b) Corresponding effect-damage function, where EDP_{local} is the ecosystem damage potential on the local scale (cf., [15]).

Tab. 7-7: Local ecosystem damage potential (EDP) calculated with the linear (equation [12]) and nonlinear model (equations [13] and [15]) for the regions Swiss Plateau (pla) and Central Alps (c-alps). EDPs are given for specific land use types, Swiss regions, 8 types of intensity of land use, and the 3 Ecoinvent types.

CORINE+ ID-No.	Type	S_{occ} for A_{st} (0.01 ha)	$S_{occ}/S_{ref\text{-}pla}$	$S_{occ}/S_{ref\_c\text{-}alp}$	$EDP_{local\text{-}lin\text{-}pla}$	$EDP_{local\text{-}lin\text{-}c\text{-}alp}$	$EDP_{local\text{-}nonl\text{-}pla}$	$EDP_{local\text{-}nonl\text{-}c\text{-}alp}$
Specific land use types								
10	Built up land	1*	0.03	0.02	0.97	0.98	0.99	1.06
2112	Integrated arable	7	0.18	0.14	0.82	0.86	0.47	0.53
142	Sport facilities	7	0.17	0.13	0.83	0.87	0.48	0.54
111	Continuous urban	8	0.20	0.15	0.80	0.85	0.44	0.50
21141	Fibre/energy crops (kenaf)	10	0.25	0.20	0.75	0.80	0.38	0.44
2111	Conventional arable	10	0.26	0.21	0.74	0.79	0.36	0.43
21142	Fibre/energy crops (hemp)	11	0.27	0.21	0.73	0.79	0.36	0.42
21143	Fibre/energy crops (chinese reed)	15	0.37	0.29	0.63	0.71	0.27	0.33
2311	Intensive meadow	17	0.42	0.33	0.58	0.67	0.24	0.30
322	Heathland	18	0.44	0.35	0.56	0.65	0.22	0.28
412	Peatbog	19	0.47	0.37	0.53	0.63	0.21	0.27
2312	Less intensive meadow	19	0.49	0.38	0.51	0.62	0.19	0.26
112	Discontinuous urban	22	0.55	0.44	0.45	0.56	0.16	0.22
3112	Semi-natural broad-leafed forest (arid)	23	0.57	0.45	0.43	0.55	0.15	0.22
2212	Organic orchard	23	0.59	0.47	0.41	0.53	0.14	0.21
3112	Semi-natural broad-leafed forest (moist)	24	0.59	0.47	0.41	0.53	0.14	0.20
1225	Rail fallow	24	0.60	0.47	0.40	0.53	0.14	0.20
1212	Industrial area part, with vegetation	24	0.61	0.48	0.39	0.52	0.13	0.20
114	Rural settlement	25	0.62	0.49	0.38	0.51	0.13	0.19
2113	Organic arable	26	0.65	0.51	0.35	0.49	0.12	0.18
141	Green urban	29	0.73	0.58	0.27	0.42	0.08	0.15

1224	Rail embankments	32	0.80	0.63	0.20	0.37	0.06	0.13
134	Mining fallow	38	0.96	0.75	0.04	0.25	0.01	0.08
321	Natural grassland	39	0.98	0.77	0.02	0.23	0.01	0.07
125	Industrial fallow	40	1.01	0.80	-0.01	0.20	0.00	0.06
113	Urban fallow	40	1.00	0.79	0.00	0.21	0.00	0.06
2115	Agricultural fallow	43	1.09	0.86	-0.09	0.14	-0.02	0.04
325	Hedgerows	44	1.12	0.89	-0.12	0.11	-0.03	0.03
2313	Organic meadow	45	1.14	0.90	-0.14	0.10	-0.04	0.03
314	Forest edge	48	1.20	0.95	-0.20	0.05	-0.05	0.01
245	Agricultural fallow with hedge-rows	53	1.34	1.05	-0.34	-0.05	-0.08	-0.01
Regions in Switzerland								
	Lakes	1						
	Mountains above timberline	27						
	Plateau	40						
	Jura	44						
	Alps	50						
Types of intensity								
	Agri_hi	13	0.32	0.25	0.68	0.75	0.31	0.37
	Forest_hi	13	0.33	0.26	0.67	0.74	0.30	0.36
	Artificial_hi	18	0.46	0.36	0.54	0.64	0.21	0.28
	Forest_li	23	0.59	0.46	0.41	0.54	0.14	0.21
	Non-use	27	0.69	0.54	0.31	0.46	0.10	0.17
	Artificial_li	29	0.74	0.59	0.26	0.41	0.08	0.14
	Agri_li	40	1.02	0.81	-0.02	0.19	-0.01	0.06
Ecoinvent type								
	IV	18	0.46	0.36	0.54	0.64	0.21	0.28
	II	27	0.69	0.54	0.31	0.46	0.10	0.17
	III	32	0.80	0.63	0.20	0.37	0.06	0.12

where S_{occ} for A_{st} 0.01 ha: species number for the specific land use type for the standardised area Ast 0.01 ha; $S_{occ}/S_{ref\text{-}pla}$ and $S_{occ}/S_{ref\text{-}c\text{-}alp}$: relative species number (where the regional number of species in pla and c-alp is the reference); pla: Swiss Plateau; c-alp: Central Alps; EDP: ecosystem damage potential; lin: linear model; nonl: nonlinear model.

* without empirical basis: one species was assumed for built up land.

7.6 *Regional characterisation factor: ecosystem damage potentials for species diversity ($EDP_{regional}$)*

In the following chapters the regional characterisation factors for ecosystem damage potential ($EDP_{regional}$) are elaborated. The effect analysis was done for Switzerland serving as the study unit (chapter 7.6.1, pp. 97). In this part the regional intensity of land use and the number of species potentially lost was assessed using a regression analysis (chapter 7.6.2, pp. 99). The results for the regional regression analysis can be viewed in chapter 7.6.3 (pp. 109). The proposal in order to calculate the damage potential based on results of the regional regression analysis can be found in chapter 7.6.4 (pp. 120). The computed results for the $EDP_{regional}$ are given in chapter 7.6.5 (pp. 122).

7.6.1 *The study unit Switzerland*

I performed the regional effect analysis for Switzerland, because it is quite diverse in terms of biogeographical regions allowing for a differentiated analysis. In addition a reason for choosing Switzerland was the accessibility of data.

It is possible to distinguish at least five large regions based on climate and topography (Fig. 7-10). These are the Jura, the Swiss Plateau, and the Northern, Central and Southern Alps. The area above timberline was analysed separately because of its specific features. For example, large portions of this region are naturally not covered with vegetation. The exact borders of the regions are drawn according to Wohlgemuth's analysis (1998), which is based on an ecological analysis of vegetation.

The Swiss Plateau is used rather intensively compared to the other regions. Performing an effect analysis for each region reveals regional differences in land use effects. Owing to a lack of historical data, I made the regional scale regression analysis for only a part of the Swiss Plateau (pla-4).

Each of the regions is divided into small polygons of about 80 km^2. The exact figures for the number of polygons and the size of their area are given in Tab. 7-8.

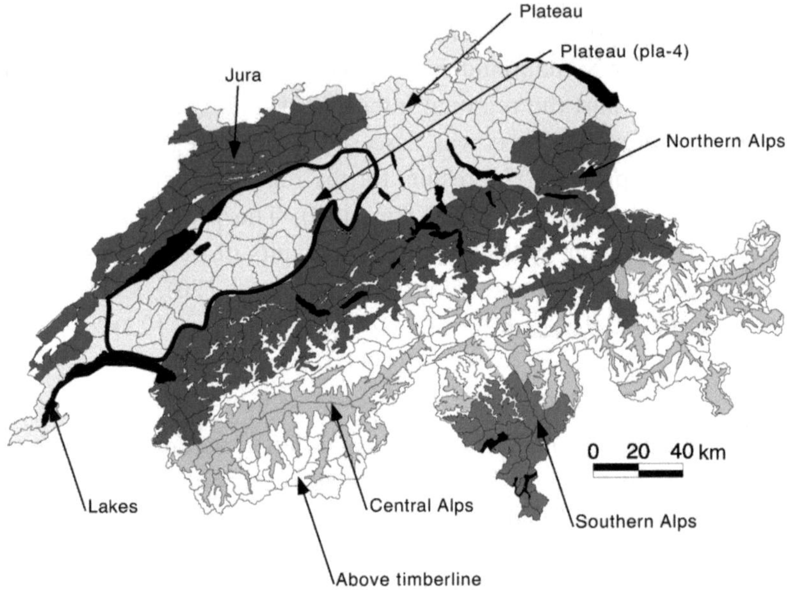

Fig. 7-10: Switzerland and its biogeographical regions: Jura, Plateau, Northern Alps, Central Alps, and Southern Alps. The bold circle marks that part of the Plateau (pla-4) which was taken into account in the effect analysis (adapted from Wohlgemuth 1996).

Tab. 7-8: Biogeographical regions of Switzerland. Italic numbers indicate that part of the Swiss Plateau (pla-4) which was considered in the regression analysis.

Region		Total area in km^2	Number of polygons	Size of polygons in km^2			
				Mean	Std Dev	Min	Max
Jura	jur	3642	43	85	27	36	172
Plateau	pla	10641	104	102	23	30	161
	pla-4	*4208*	*39*	*108*	*18*	*79*	*142*
Northern Alps	n-alp	9073	110	82	25	23	150
Central Alps	c-alp	5222	71	74	26	28	152
Southern Alps	s-alp	1449	21	69	37	11	153
Above timberline	timberl	9853	215	46	32	0.1	134
Lakes	water	1248	28	45	48	5	228
total		41128	592				

7.6.2 *Regional intensity of land use and species potentially lost*

Basic idea of assessment applying a regression analysis

In order to determine the ecological damage/benefits of each specific land use type, I assessed its potential impact on the regional species-pool applying a regression analysis. The size of assessed regions varies from 1'500 km^2 to 10'000 km^2 (Fig. 7-10). The result is a regional scale characterisation factor – the ecosystem damage potential ($EDP_{regional}$).

The basic idea behind this assessment is to relate the proportion, for every polygon of about 80 km^2, of the type of land use present to the species within it which have potentially been lost (Fig. 7-11). I used the database EDV-Flora 1.0 (Wohlgemuth 1993) to determine the potential loss of species since ~1850. I assumed that land use types vary in their contribution to the potential disappearance of species.

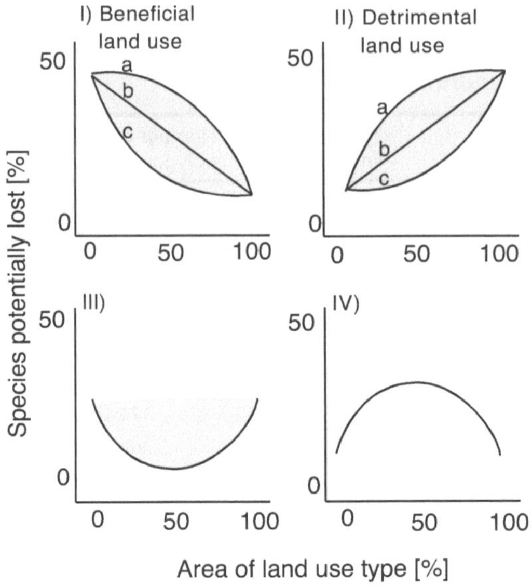

Fig. 7-11: Possible models showing the relationship between the percentages of total land characterised by a type of land use and the corresponding species potentially lost regionally since ~1850.

The following are some possible relationships between the percentage of species potentially lost and a specific land use type (measured as either absolute or relative number):

- I negative relationship: the larger the area or proportion of a type of land used the fewer the species lost (e.g., unfertilised meadow);
- II positive relationship: the larger the area or proportion of a type of land used the more the species lost (e.g., artificial area);

Considering percentages only, two more relationships are possible:

- III U-shape relationship: when the proportion of one land use type is low many species are lost, when the proportion is medium a minimum of species is lost, when the proportion is large a large number of species is lost (e.g., some forest species are missing when the proportion of forest is low, but at the turning point, where forest becomes dominant, non-forest species associated with other land use types decline);

IV inverted U-shape relationship: that is the opposite of case III;

For cases I and II percentages of species lost (S_{lost}) may change as the land area/proportion (L_i) changes either linearly,

$$S_{lost} = a + bL_i \textit{ (Fig. 7-11 Case I b, IIb)} \qquad [16]$$

exponentially,

$$S_{lost} = e^{a+bL_i} \textit{ (Fig. 7-11 Case I c, II c)} \qquad [17]$$

or logarithmically

$$S_{lost} = a + b\ln(100 - L_i) \textit{(Fig. 7-11 Case Ia)} \qquad [18]$$

$$S_{lost} = a + b\ln(L_i) \textit{(Fig. 7-11 Case IIa)} \qquad [19]$$

I assumed that the negative impact will grow exponentially as the area of detrimental land use types grows (case II c) and that negative impact will decline exponentially as the beneficial land use types increases (case I c). The reasoning for that assumption is as follows. It is likely that an additional area of a detrimental land use type triggers a large additional potential loss of species in a hypothetical region *Malus* which is anyway intensively used and thus, shows only a low proportion of beneficial land use in terms of species diversity. In contrast, the same additional increase in detrimental land use is expected to result in a small number of species lost in addition in a region *Bonus* with a large proportion of beneficial land use. The same principle is expected for changes in beneficial land use types. The same changes have a larger impact in a region *Malus* compared to a region *Bonus*.

For these reasons, I used the exponential function for the regression analysis. In order to compare results I got using the nonlinear relationship, I applied the linear relationship as well.

The regression was performed separately for each of the 5 biogeographical regions. The independent variable in the regression analysis was the proportion of different types of intensity of land use in Switzerland. The dependent variable was the percentage of species potentially lost on the regional scale.

Since the proportions of different land use types in a given region are interdependent, I also calculated the correlation between the proportions of land use types. It is possible that this dependency negatively affects the regression analysis.

Independent variable: Proportion of different types of intensity of land use based on the modified area statistics of Switzerland

In order to determine the current intensity of land use in Switzerland for each polygon in Fig. 7-10, I used the Swiss Area Statistics (BFS 1992) and the Hegg atlas (Hegg et al. 1993). The unmodified version of the Swiss Area Statistics could not be used in this investigation, because it does not distinguish between (i) high and low-intensity closed forest and (ii) high and low-intensity arable land and meadow.

The eight types of intensity of land use I took included in the regression analysis are shown in Tab. 7-9. The first column gives the 26 land use types found in the official Swiss Area Statistics (BFS 1992). The parenthesis of the first column indicate the refinement of the area statistic, i.e., closed forest and agricultural area (i.e., arable land and meadow) are split using information from Tab. 7-10 below. This refinement results in 29 land use types. The second column gives the ID-numbers of the 69 land use types assigned in the Swiss Area Statistics (BFS 1992). In the last column the aggregation into high-intensity and low-intensity land use types is shown. This aggregation results in eight types of intensity of land use:

- High-intensity forest (forest_hi)

- Low-intensity forest (forest_li)
- High-intensity agricultural area (agri_hi)
- Low-intensity agricultural area (agri_li)
- Lakes and rivers (lakes)
- Terrestrial non-use area (non-use)
- High-intensity artificial surfaces (artificial_hi)
- Low-intensity artificial surfaces (artificial_li)

The heterogeneity of the type *non-use* is very large since non-productive vegetation is in general very rich in terms of species diversity and area without vegetation very species poor. However, those areas without vegetation are mostly above the timberline, which is a region I excluded from my analysis. Therefore, in the type of intensity *non-use* the non-productive vegetation dominates in regions included in the analysis.

Tab. 7-9: Refinement of the Swiss Area Statistics (BFS) and aggregation into the types of intensity considered in the regression analysis, where hi is high-intensity and li low-intensity.

Land use type of Swiss Area Statistics, BFS *(refinement using Hegg Atlas)*	BFS ID-no	Aggregated into hi and li
Forest		
Closed forest *(high-intensity closed forest)*	10, 11, 14	forest_hi
Closed forest *(low-intensity closed forest)*	"	forest_li
Open forest	12, 13	forest_li
Shrub forest	15	forest_li
Hedgerows, cluster of trees	17, 18, 19	forest_li
Agricultural area		
Vineyards	71, 72	agri_hi
Intensive fruit trees	75, 76	agri_hi
Extensive fruit trees	77	agri_li
Horticulture	78	agri_hi
Arable land and meadow *(Arable land)*	81, 82	agri_hi
Arable land and meadow *(Intensive meadow)*	"	agri_hi

Land use type of Swiss Area Statistics, BFS *(refinement using Hegg Atlas)*	BFS ID-no	Aggregated into hi and li
Arable land and meadow *(Organic meadow)*	"	agri_li
Extensive vineyards	73	agri_li
Meadows (Heimweiden)	83, 84	agri_li
Meadows (Maiensässen, Heualpen, Bergwiesen)	85	agri_li
Pastures in the Alps and the Jura	86, 87, 88, 89	agri_li
Non-productive area		
Lakes	91	lakes
Running water with embankment	62, 92	lakes
Non-productive vegetation	16, 95, 96, 97	non-use
Area without vegetation	99	non-use
Artificial surfaces		
Buildings – built up part	25, 26, 27, 28, 29	artificial_hi
Buildings – part with vegetation	45, 46, 47, 48, 49	artificial_li
Industrial plants – built up part	21	artificial_hi
Industrial plants – part with vegetation	41	artificial_li
Special artificial surfaces	20, 24, 61, 62, 63, 65, 66	artificial_hi
Sport, leisure and green urban areas	23, 51, 52, 53, 54, 56, 59	artificial_li
Road networks	31, 32, 33, 34, 68	artificial_hi
Rail networks	35, 36, 67	artificial_hi
Airports	37, 38	artificial_hi

Tab. 7-10: Information on splitting closed forest and agricultural area (i.e., arable land and meadow) into high- and low-intensity land use types based on the Atlas of Hegg et al. (1993)

Swiss area statistics +	Hegg (Hegg ID- no.)
Forest	
Closed forest, high-intensity	Plantations of robinia, birch, beech, chestnut (467, 468, 469, 457, 489, 567, 568)
	Plantations of spruce, pine, larch (379, 389, 456)
	General plantations (378)
Closed forest, low-intensity	Beech forest, xerothermic broad-leafed forest, fen alder wood (148, 267, 268, 269, 278, 279, 289, 345, 346, 347, 348, 349,

Swiss area statistics +	Hegg (Hegg ID- no.)
	356, 358, 359, 367)
	Silver fir, spruce, larch-arolla pine wood and mountain pine wood (149, 156, 157, 158, 159, 179, 189, 234, 235, 236, 237, 238, 239, 245, 246, 247)
	Mesophile oak mixed wood, lime and sycamore-Ash wood, Oak-Birch wood (249, 357)
Agricultural area	
Arable land (l_ch2.05)	Arable land, vegetable (678, 689)
Intensive meadow (l_ch2.06)	Fertilised meadows in valleys (57)
Extensive meadow (l_ch2.07)	Tall-herb communities, wet litter meadows (68, 69)

Data in the Hegg Atlas was used to split each of the agricultural and forest areas into a high-intensity part and low-intensity part. To do this, the data from the Hegg Atlas was processed as follows:

- The vegetation types in the Hegg Atlas were attributed to the extended Swiss area statistics according to Tab. 7-10;
- The Hegg Atlas provides information about the occurrence of each vegetation type on every km^2. Thereby, the area of each type of vegetation was measured in 4 ordinal categories of size (fragment, small, medium, and large). For each km^2 information about type of vegetation and its size is available. Next, the count of each size category within one km^2 was determined for high-intensity closed forest, low-intensity closed forest, arable land, intensive meadow and extensive meadow. In order to obtain cardinal figures I used the following calculation:
  $$count(land\_use\_type_i) = 1 * counts_i(fragments) + 10 * counts_i(small) + 100 * counts_i(medium) + 1000 * counts_i(large)$$
- The result is a figure for the occurrence of land use type *i* for each one-km^2 cell of the grid of Switzerland.

- To get a percentage for the occurrence of land use type i I used the following calculation:

$$land\_use\_type_i * 100[\%] = \frac{count(land\_use\_type_i)}{\sum_{i=1}^{n} count(land\_use\_type_i)}$$

- In the last step, I calculated the mean proportion of land use type i for each EDV-Flora polygon by spatial intersection using the Geographical Information System (ArcView by ESRI).

Data for land use 1: The Swiss Area Statistics (BFS 1992)

The Swiss Area Statistics gives information about which of 69 land use types is present on every one-hectare grid cell for the whole country. To do this a hectare grid was put on top of aerial photos and the land use type indicated in the lower left corner was assessed by the researchers. Debatable classifications were verified by a ground survey (BFS 1992).

Data are available for the Area Statistics from the years 1979-1985 (i.e., aerial photos are taken between these years). More recent statistics (1992-1997) are not ready yet for total Switzerland. Moreover the biodiversity data for Switzerland that I used were taken during the period of the first area statistics. For these reasons, I did not take the new area statistics of 1992-1997 into consideration.

Data for land use 2: The Hegg Atlas (Hegg et al. 1993)

Beyond the Area Statistics additional information was provided by the Hegg Atlas of Threatened Vegetation Types. In this atlas, the occurrence of 118 threatened vegetation types is mapped on a one-km^2 grid. Also, the presence of 21 types of vegetation which are based on human activities are considered (e.g., tree plantations). The authors made distinctions between the vegetation types based on topographical and geological maps as well as aerial photos. Ground surveys were also undertaken. The atlas was created in the years from 1974 to 1981.

Dependent variable: Species potentially lost on the regional scale

I used the database EDV-Flora 1.0 (Wohlgemuth 1993) to determine the potential loss of species in Switzerland. In addition to the 1967-1978 ground survey data this database includes data on plant species, which are known from literature and herbaria, but could not be found in the rather thorough 1967-1978 ground survey (Fig. 7-12). This disparity represents the number of species potentially lost. The historical data was compiled between 1850 and 1957. Therefore, it can be concluded that the species were lost between 1850 and the 1967-1978 ground survey.

I calculated the species number potentially lost S_{lost} as

$$S_{lost} = \frac{S_{\sim 1850} - S_{\sim 1978}}{S_{\sim 1850}} = 1 - \frac{S_{\sim 1978}}{S_{\sim 1850}} \qquad [20]$$

where $S_{\sim 1850}$ is the species number found between 1850 and 1957 and $S_{\sim 1978}$ the species number found more recently in the ground survey 1967-1978.

The historical data does not cover all of Switzerland. Therefore, I included only parts of the plateau in my analysis (Fig. 7-10, p. 98). After the publication of EDV-Flora 1.0 botanists were able to find additional species, which had been considered as lost. However, I did not use the most recent data, because the new survey is rather unsystematic and only concentrates on some areas of Switzerland. The historical sources, however, also vary in quality and contribute to the overall uncertainty of the analysis.

The basic assumption behind the regression analysis is that different land use types (e.g., high-intensity artificial area versus low-intensity agriculture) contribute to the potentially disappearance of species differently.

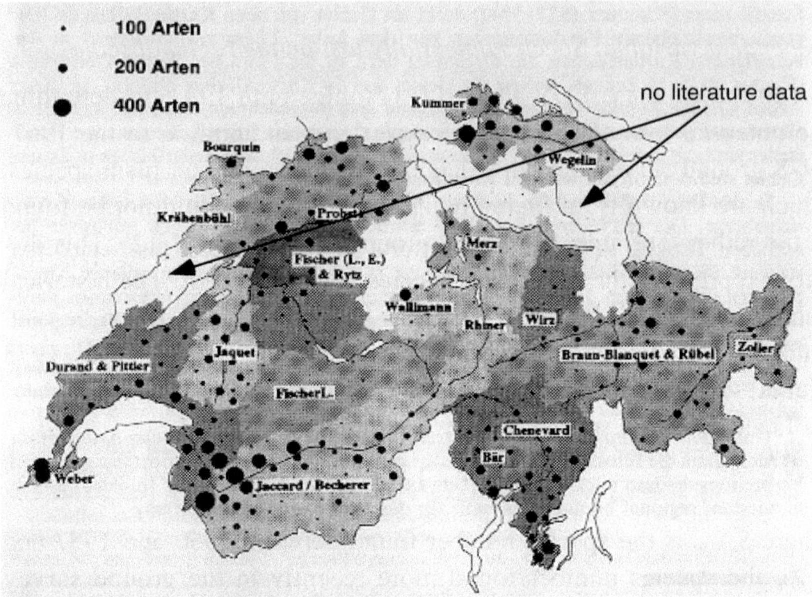

Fig. 7-12: For each polygon of Fig. 7-10, this map shows the species number (=Arten) which are known from literature, but could not be found in the ground survey. Names for historical written sources are given (taken and adapted from Wohlgemuth 1993) in the middle of its study area.

Data for regional biodiversity: The plant species richness of Switzerland, Welten and Sutter Atlas (Wohlgemuth 1993)

The source for information about the plant species richness was the atlas by Welten and Sutter (1982) which was digitised as EDV-Flora 1.0 and put into a Geographical Information System by Wohlgemuth (1993). The atlas includes data from an extensive and rather systematic ground survey. About 200 botanists did an inventory of the vascular plant species throughout Switzerland between 1967 and 1978. To do this, the surface of the country was divided into about 600 polygons (Fig. 7-10) of about 80 km^2. For each polygon, a list of species type and number is available. Historical data on plant species richness was included in addition.

7.6.3 Results for regional effects

In order to perform a regression analysis between the types of intensity of land use (independent variable) and the species potentially lost since ~1850 (dependent variable), I calculated the variables for each of the ca. 592 polygons of Switzerland. The regression analysis was made for each Swiss region (see map in Fig. 7-10). Due to a lack of historical data for some parts of the Swiss Plateau I took it into account only partially (pla-4). I distinguished the area above the timberline from that below the timberline in the Swiss Alps and performed the analysis for each type separately. This is because the area above timberline differs from the other area below in terms of species richness. Although large areas are not used above the timberline species richness is rather low because of extreme abiotic conditions. A simple regression analysis was done for all the species and those that are threatened separately. In addition a multiple regression analysis was performed.

Independent variable used for the regression analysis: Proportions types of intensity of land use in Switzerland

The proportions of the types of intensity of land use in each Swiss region and the average for the whole country are shown in Tab. 7-11. The more detailed results are shown in appendix 2. The results vary considerably from region to region. The Swiss Plateau is used the most intensively and has large proportion of high-intensity agriculture and artificial area. In contrast, the proportions of low-intensity forestry and agriculture in alpine regions are large and their proportions of artificial area is small. The surprisingly high proportion of high-intensity forestry in the Southern Alps is due to the fact that I assigned the Castanea forests to the high-intensity forests. From a botanical point of view, however, these forests, when uncultivated, can be very rich.

Since the types of intensity are assumed not to be independent from each other its correlation was calculated (Tab. 7-12). High-intensity agriculture and low-intensity forestry for example correlates negatively. Artificial area and high-intensity agriculture are positively correlated. Generally speaking, the high-intensity types occur in combination with one another. At the same time the low-intensity types are found to be correlated.

Tab. 7-11: Proportions of types of intensity of land use for Swiss regions and the country average.

Type	jur	pla-4	n-alp	c-alp	s-alp	above timberl	surface water	Swiss ø
forest_hi	7.0	8.2	3.9	3.1	22.8	0.2	0.2	5.1
forest_li	41.5	17.6	39.6	45.5	46.1	6.7	2.6	26.1
agri_hi	20.4	52.8	14.6	7.2	7.0	0.3	0.7	18.7
agri_li	23.9	9.3	27.2	20.3	5.9	30.1	1.1	20.5
lakes	0.5	1.3	2.4	2.2	2.2	0.5	94.1	4.2
non-use	0.4	0.3	7.9	18.3	6.8	62.0	0.5	19.4
artificial_hi	3.5	5.8	2.7	2.2	4.8	0.2	0.4	3.3
artificial_li	2.9	4.8	1.8	1.1	4.5	0.0	0.5	2.6
total	100.0	100.0	100.0	100.0	100.0	100.0	100.0	100.0

Type of intensity: forest_hi: high-intensity forest, forest_li: low-intensity forest, agri_hi: high-intensity agriculture, agri_li: low-intensity agriculture, lakes: lakes and rivers, non-use, artificial_hi: high-intensity artificial area, artificial_li: low-intensity artificial area, Swiss ø: Swiss average; **Region:** jur: Jura, pla-4: part of Plateau, n-alp: Northern Alps, c-alp: Central Alps, s-alp: Southern Alps, above timberl: area above the timberline, water: lakes;

*Tab. 7-12: Correlation between the percentages of types of intensity of land use in Switzerland below timberline (n=350) where * means p ≤ 5% and ** means p ≤ 1%.*

	for-est_hi	forest_li	agri_hi	agri_li	non-use	lakes	artif._hi	artif._li
forest_hi	1.00							
forest_li	-0.16	1.00						
agri_hi	0.21	-0.81**	1.00					
agri_li	-0.52	0.33**	-0.57**	1.00				
non-use	-0.41	0.46**	-0.62**	0.26**	1.00			
lakes	-0.07	-0.10*	-0.12*	-0.08	0.05	1.00		
artificial_hi	0.32	-0.58**	0.53**	-0.53**	-0.47**	-0.06	1.00	
artificial_li	0.32	-0.53**	0.46**	-0.51**	-0.45**	-0.05	0.95**	1.00

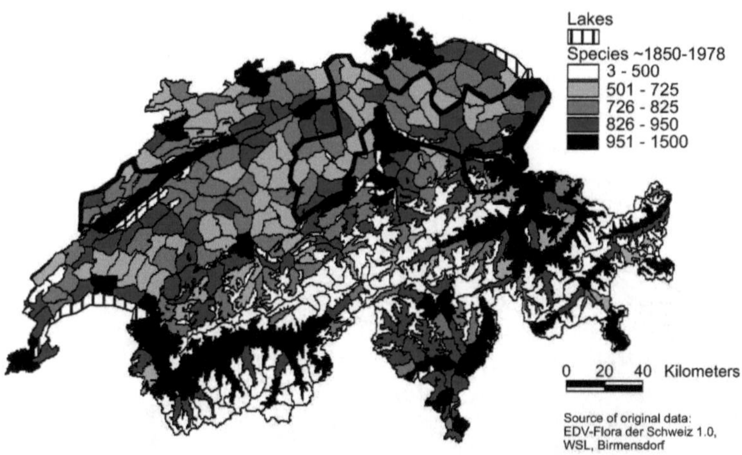

Fig. 7-13: Number of plant species found in the 1967-1978 ground survey plus additional plant species according to the historical data on species occurrence. For areas which are circled with a bold line historical data only were available from herbaria but not from literature.

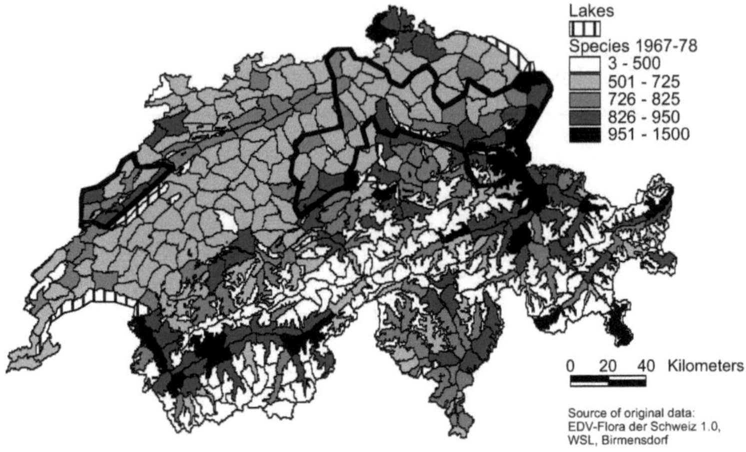

Fig. 7-14: Number of plant species according to the 1967-1978 ground survey.

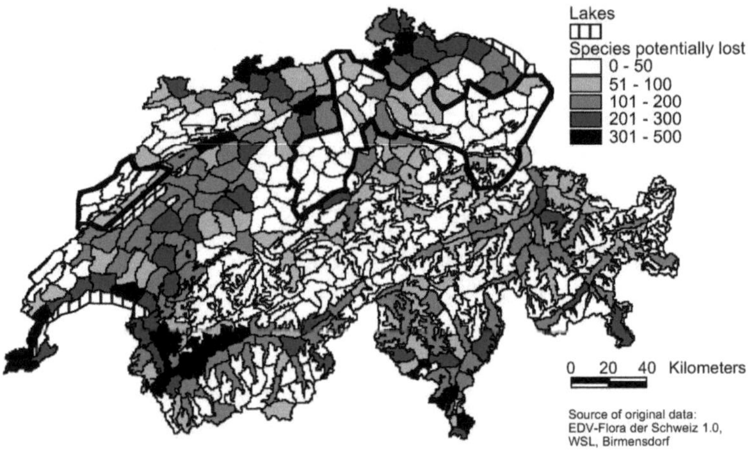

Fig. 7-15: The number of potentially lost plant species refers to the species known from historical sources (literature and herbaria), but were not found again in the 1967-1978 ground survey.

Dependent variable used for the regression analysis: Species potentially lost in Switzerland

The dependent variable used for the regression analysis, is the number of plant species potentially lost in Switzerland. Number of plant species found in the 1967-1978 ground survey together with the historical data on species occurrence are seen in Fig. 7-13. The Fig. 7-14 gives the number of plant species found in the ground survey between 1967 and 1978. Comparing both figures it is obvious that the picture of 1967-1978 ground survey is more homogeneous and that species richness has decreased. Fig. 7-15 shows the number of potentially lost plant species in each polygon. The number refers to species which are known from historical sources (literature and herbaria), but could not be found again in the rather extensive 1967-1978 ground survey. In some polygons there are up to 500 species potentially lost, which is surprisingly high for an area of about 80 km^2. Apparently, the species lost do not have a random pattern. Hot spots of loss are at the lakeshores, the south-western part of the Swiss Plateau, and the valleys of the Alps. In these regions land use became more intensive during the 20th century. Especially, silviculture, agriculture and artificial land use caused this intensification.

Results of the regression analysis

The regression analysis results in figures for the functional relationship between the intensity of land use and the species potentially lost on the regional scale.

To explain the results, which are given in detail in appendix 3, I have chosen to use the types *low-intensity agriculture* and *high-intensity artificial area* in the Jura as examples. The absolute number of species potentially lost is calculated as the difference between the total species number and the species found in the 1967-1978 ground survey (Fig. 7-16, top graph). There was no polygon in Switzerland where an increase in species numbers was observed according to EDV Flora 1.0. This does not mean that in the species pool of one polygon no new species have occurred, rather than no polygon with a net increase taking species loss and species gain into account was observed. In addition, neophytes are not considered in that database.

The percentage of species potentially lost (S_{lost}) is negatively correlated with the percentage of low-intensity agriculture (L_{agri_li}) where $R^2 = 0.3$ and $S_{lost} = 15.37\text{-}0.27L_{agri\_li}$ (Fig. 7-16, bottom graph). One possible interpretation of this result is that low-intensity agriculture works as a "protection" against species loss. The other example (Fig. 7-17, bottom graph) shows that S_{lost} is positively correlated ($R^2 = 0.46$) with the proportion of high-intensity artificial land use ($S_{lost} = 0.35 + 2.57L_{agri\_li}$). An increase in that land use type of one percent triggers a 2.57 percent loss of species.

The *individual regression analysis* was calculated twice: One time for all the species and, another time for those species that are threatened. The difference in the results is shown in Fig. 7-18 where the slope parameters *b* (cf., equation [16]) resulting from significant regressions of the linear type are exemplary plotted. This parameter relates to the steepness of the regression function and is a measure for the marginal change of the species potentially lost in % when a change of 1% of the area of the corresponding land use type occurs. This means a 1% increase type *i* leads to a b% increase in species lost.

All the species and those that are threatened on a national level show in general the same response to land use changes. However, the response of threatened species is more sensitive.

Regression plots for every type of intensity and all regions are given in appendix 3. The values for the parameters of the linear (equation [16]) and nonlinear regression functions (equation [17]) can be found in appendix 3 as well. Please note that the correlation coefficient varies considerably, that is, the explained variance in the proportion of a specific land use type can be quite low, even for significant results.

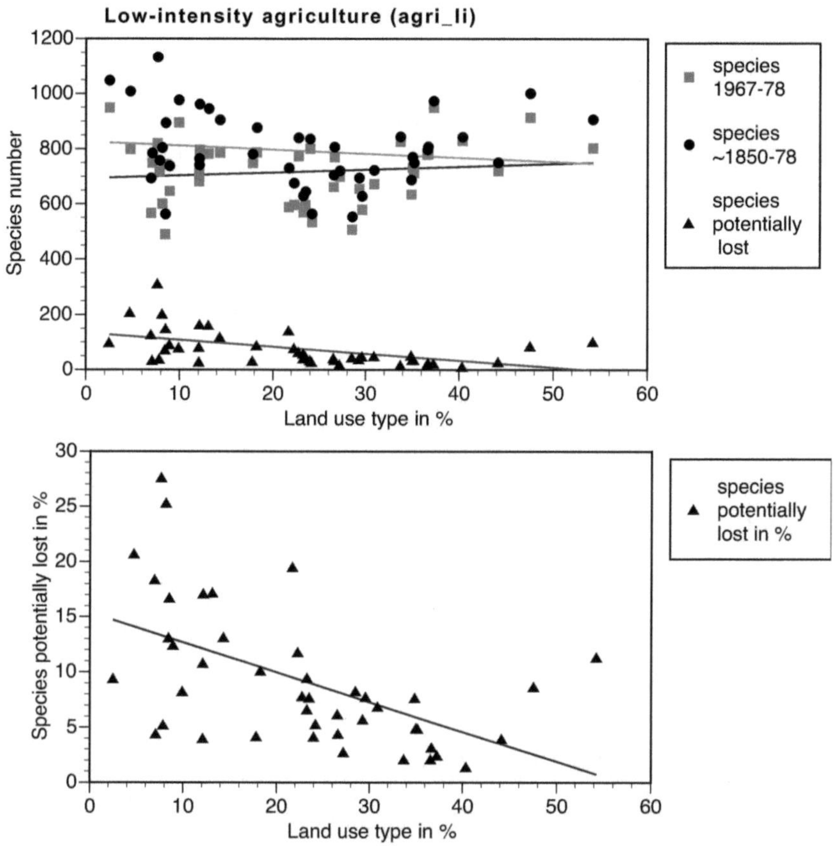

Fig. 7-16: Linear regression of the proportion of low-intensity agricultural land (independent variable) and species number (dependent variable) in the Jura.

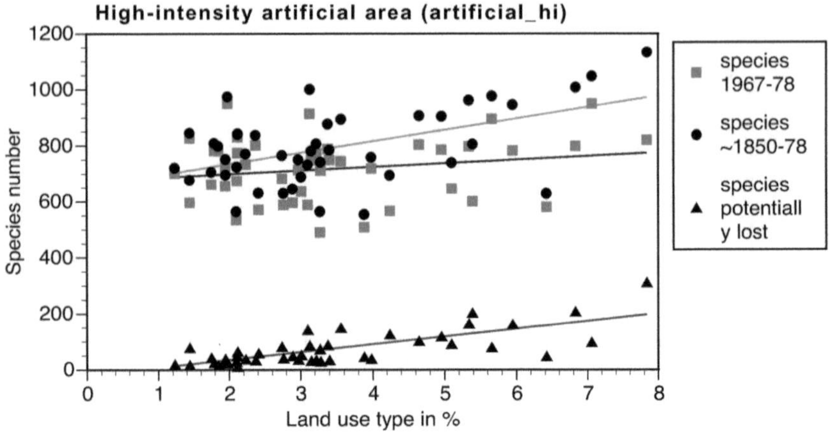

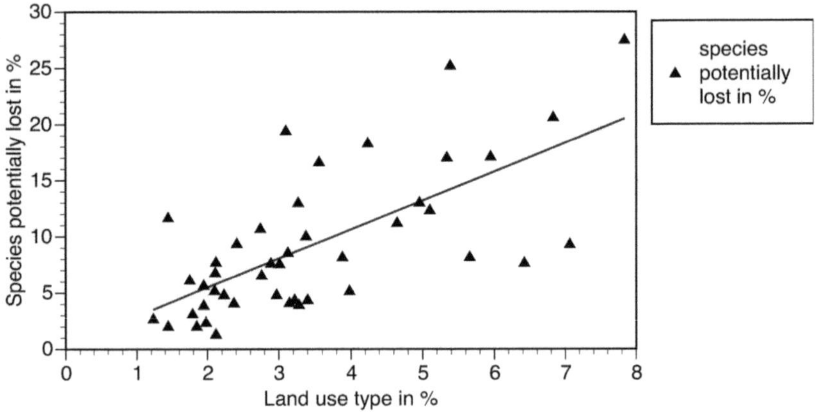

Fig. 7-17: Linear regression for the proportion of high-intensity artificial land (independent variable) and species number (dependent variable) in the Jura.

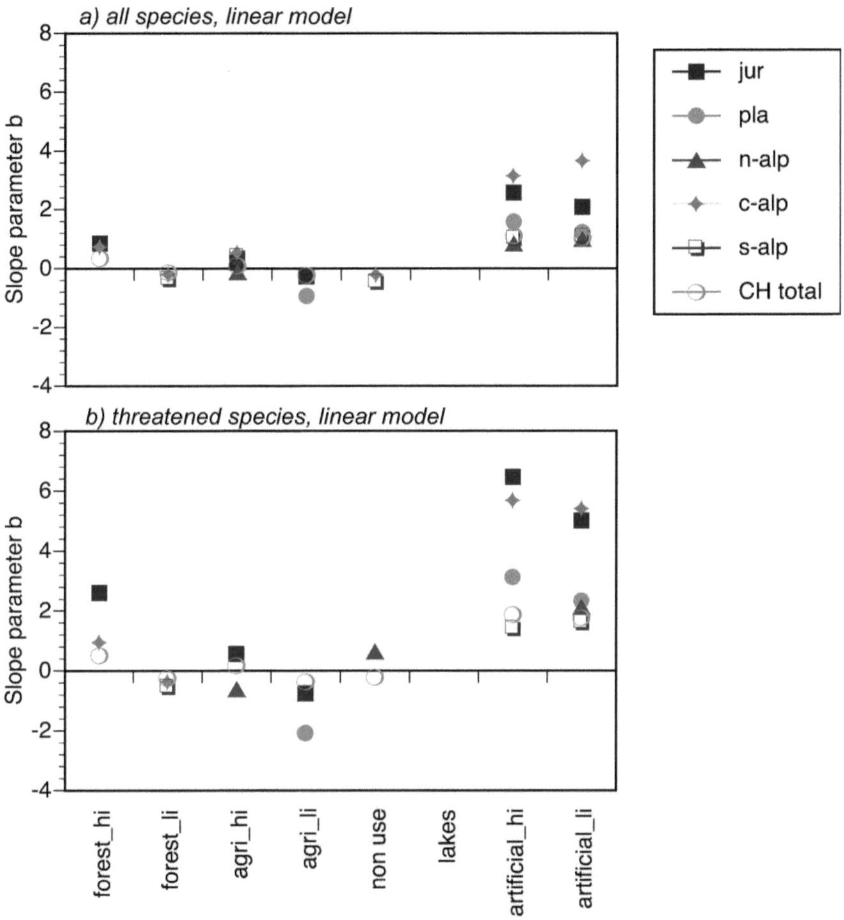

Fig. 7-18: Percent change in species potentially lost (equals parameter b in equation [16]) for every 1% change in the proportion of a specific type of intensity for both all of the species (graph a) and only those species that area threatened (graph b). Only values for linear regressions which were significant were considered.

In addition to the analysis, in which I performed individual independent regressions, I created also a *multiple regression model* (Tab. 7-13).[11] In this model the independent variables are the land use types, the area of polygons and the west-east co-ordinates and north-south co-ordinates of the polygons in space. Only those independent variables, which contribute to the total fit remain in the model. For example the regression model for the Jura for all species potentially lost (S_{lost}) was determined as:

$$S_{lost} = -2.30 + 0.53\, forest\_hi\ [\%] + 2.25\, artificial\_hi\ [\%].$$

This means that for this region S_{lost} is a function of *forest_hi* and *artificial_hi*, which are the independent variables in the multiple regres-

11 The general form of such a model is $y = constant + ax_1 + bx_2 + cx_3$ ..., where y is the dependent variable, x_i are the independent variables and a,b,c... are parameters of the function. The regression coefficient R^2 refers to the explained variance of the dependent variable and is calculated as:

$$R^2 = \frac{explained\_variance}{total\_variance} = \frac{\sum_{k=1}^{N}(\hat{y}_k - \bar{y})^2}{\sum_{k=1}^{N}(y_k - \bar{y})^2}$$

where k refers to one observation out of N observation in total, y_k is the individual value of one observation, $\bar{y}$ is the arithmetic mean of all y-values, $\hat{y}_k$ is the value calculated for an observed x_k using the regression function. The adjusted R^2 takes also the degree of freedom (DF= N -V - 1, where V is number of independent variables), into consideration. The F-statistics is used to test whether or not dependency is given between explanatory variables and presumable dependent variables where the null hypothesis is that the slope parameter is zero. The value for the F-statistics is calculated as

$$F - statistics = \frac{\frac{explained\_variance}{V}}{\frac{unexplained\_variance}{N - V - 1}} = \frac{\frac{R^2}{V}}{\frac{1 - R^2}{DF}}$$

which is then used to calculate the significance level α for a specific regression model. When α was chosen to be smaller than 10%, the null hypothesis can be rejected with a probability of confidence of 90% and thus dependency is given based on statistical reasoning. The value SignF refers to the level of significance calculated for the individual model (cf., Backhaus et al. 1996).

sion model. The inclusion or deletion of any variables would result in a decrease of the regression coefficient R^2. However, this does not mean that excluded variables are meaningless. For example high-intensity agriculture is not included in any multiple model, however, on the basis of the results of the individual regressions it can be stated that an increase in that type of intensity triggers in some regions an increase in the species lost.

The results show that for selected independent variables the explained variance for species potentially lost ranges from 20% to 54%. The other part is unexplained variance and is a result of other independent variables not considered here. Overall it can be concluded that the results of the individual regression models correspond with those of the multiple regression models.

In general, the correlation between all the species and threatened species which have potentially been lost is quite high (Fig. 7-19). It is $R^2 = 0.85$ for the absolute number of species and $R^2 = 0.62$ for the relative numbers of species. The relative results are quite remarkable. If the percentage of all species lost is low (~10%), the number of threatened species lost can vary between 20% and 80%.

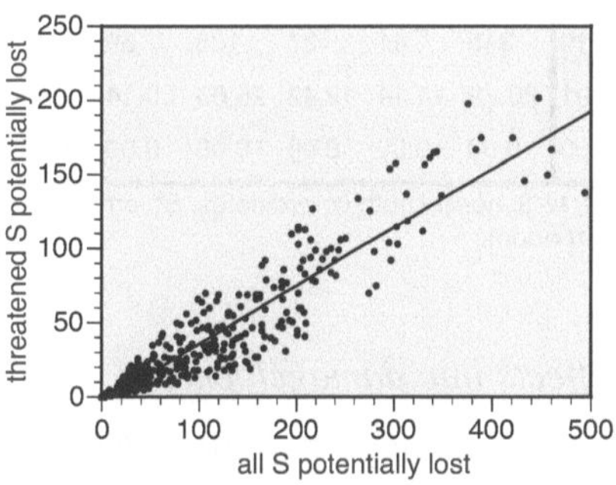

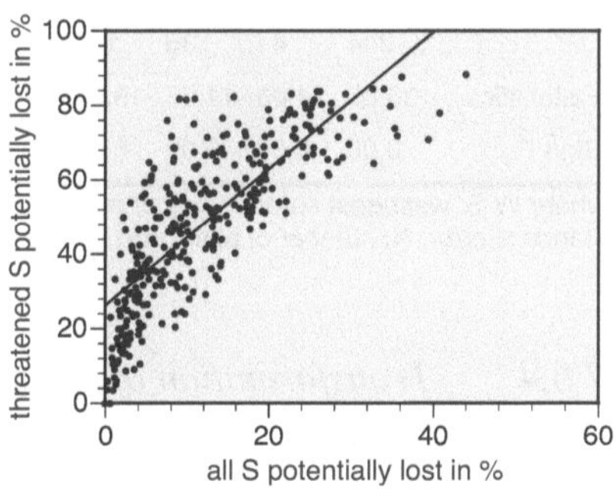

Fig. 7-19: Correlation between all the species potentially lost and the threatened species potentially lost. Each dot represents one polygon of Switzerland. Absolute numbers (left graph) and relative numbers (right graph) are given.

Tab. 7-13: Multiple linear regression model for Swiss regions. Each column refers to an individual model.

	all species potentially lost						**threatened species potentially lost**					
	CH	jur	pla-4	n-alp	c-alp	s-alp	CH	jur	pla-4	n-alp	c-alp	s-alp
constant	26.83	-2.30	16.07	25.20	19.11	11.71	90.10	17.62	78.44	103.4	33.64	42.16
forest_hi		0.53					0.44	1.85				
forest_li												
agri_hi												
agri_li	-0.13		-0.65						-2.08			
non-use							0.52					
lakes												
artificial_hi	1.00	2.25	1.05				1.93	5.36			5.67	
artificial_li				0.88	2.43	1.09						1.60
W-E	-0.12						-0.86					
N-S	-0.62			-1.02	-0.87					-3.27		
Area of polygon	0.00				0.00							
R^2	0.33	0.54	0.40	0.27	0.47	0.39	0.19	0.41	0.25	0.20	0.23	0.24
Adjust. R^2	0.32	0.52	0.36	0.25	0.44	0.35	0.16	0.38	0.23	0.19	0.21	0.21
St. error	7.24	4.38	6.23	5.95	5.35	6.34	0.18	15.23	16.64	19.99	13.75	12.96
N	349	43	38	109	70	20	349	43	38	109	70	20
DF	344	41	36	107	67	19	345	41	37	108	69	19
F statistics	33.88	24.03	11.86	19.29	19.62	11.91	20.79	14.14	12.42	26.63	20.14	6.16
Sign F	0.00	0.00	0.00	0.00	0.00	0.00	0.00	0.00	0.00	0.00	0.00	0.02

where W-E: west-east co-ordinates of an individual plot, N-S: north-south co-ordinates, St. error: standard error, N: number of plots, and DF: Degree of freedom.

7.6.4 Transformation of regional effects into damage/benefits

To calculate the regional scale ecosystem damage potential ($EDP_{regional}$), I used the results of the regression analysis. There the proportion of a land use type was correlated with the species potentially lost (Fig. 7-20a). In chapter 7.6.2 different relationships were proposed and compared.

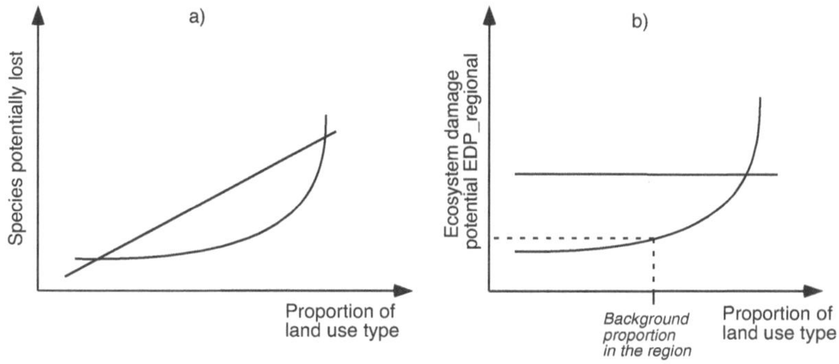

Fig. 7-20: a) Possible functions for the proportion of land use type (L_i) and potentially lost species in the region, which are considered in the regression analysis (chapter 7.6.2). b) Corresponding effect-damage functions, which equal the first derivative of curves given in graph a.

I applied a marginal approach on the regional scale. I assume that the land use activities being assessed in a specific LCA themselves do not change the background proportion of a land use type in the region. This means that in Fig. 7-20b the slope of curves at the point of the background proportion is taken as the indicator for the damage.

Option 1: Linear damage function

The number of species potentially lost S_{lost} is a function of the proportion of land use type *i* (L_i):

$$S_{lost} = a + bL_i \qquad \textit{[16] page 101}$$

where the parameter *a* and *b* are assessed with the regression analysis.

The ecosystem damage potential ($EDP_{regional}$) can be found by the first derivative of equation [16] which is then

$$EDP_{regional} = \frac{dS_{lost}}{dL_i} = b \qquad [21]$$

Obviously, here the damage potential is independent of the background proportion of the land use type (Fig. 7-20b).

Option 2: Nonlinear damage function

In the exponential case, the function used for the correlation is

$$S_{lost} = e^{a+bL_i} \qquad [17]\ page\ 101$$

EDP is calculated as the first derivative:

$$EDP_{regional} = \frac{dS_{lost}}{dL_i} = be^{a+bL_i} \qquad [22]$$

Here the background proportion of land use type *i* (L_i) is taken into account. The marginal damage due to detrimental land use activities is assumed to increase as the proportion of this type of land use in the region increases. Likewise, the marginal damage which comes out of an increase of beneficial land use types is assumed to decrease (cf., Fig. 7-1 case I, II c). This choice has the following consequence. The use of 1 hectare for a parking lot (no species, very bad) means a high degree of damage in an already stressed region and a lower damage in a region where the background situation is good and detrimental land use is low.

7.6.5 Results for $EDP_{regional}$

To calculate the ecosystem damage potentials on the regional scale ($EDP_{regional}$), the parameters of the regression analysis were used. I only took those parameters resulting from significant regressions into account. This means that the dependency between the percentage of the type of intensity and the percentage of the species potentially lost is statistically proofed (marked with an "°" in the tables in appendix 3, pp. 203). In addition, only those with a correlation coefficient $R^2 \geq 10\%$ were included, because I assume that regressions with a lower R^2 are not meaningful at all. The regression functions considered show an R^2 between 10% and 46%. The best solution would have been to have used parameters in the damage functions (equations [21] and [22]), which are specific for each region. However, due to a lack of significant regressions for every type of intensity in each region this was not possible. Instead I determined the mean value of the parameters of the Swiss regions (Tab. 7-14, p. 124) and its standard deviation with a Monte-Carlo simulation. I then used this mean and standard deviation in the damage function. The original data are shown in appendix 3. The map of the total study area and its regions is given in Fig. 7-10 (p. 98).

In the case of the linear damage function, the marginal damage is assumed to be independent of the region or the specific proportions of types of intensity in that region. Here the mean of the parameter *b* of the regression line (Tab. 7-14, p. 124) equals the first derivative, and therefore, the damage value (cf., equation [21] and Tab. 7-15).

For the nonlinear damage function, the damage is dependent on the proportion of each land use type L [%] in the region. It is calculated according to equation [22]. As examples the results are given for the Swiss Plateau (pla) and the Central Alps (c-alp) (Tab. 7-16), because those regions are very characteristic in terms of land use intensity. The Plateau is used much more intensively than the Central Alps. Therefore, damage and benefits are more extreme in the Plateau than in the Central Alps. This implies that a 1% decrease of beneficial types of intensity of land use in the Plateau triggers a large damage compared to the same decrease in the Central Alps. Conversely, an increase in beneficial types triggers large benefits in the intensively used Plateau.

Tab. 7-14: Values of the parameters of the regional regression analysis (equations [16] and [17]) resulting from significant regressions with an $R^2 \geq 10\%$.

		Parameter *a*				Parameter *b*			
	n	Mean	Std.Dev	Min	Max	Mean	Std.Dev	Min	Max
Linear model – all species									
Agri_hi	3	9.4	1.0	5.2	12.6	0.38	0.08	0.13	0.66
Agri_li	2	19.8	1.3	15.2	23.7	-0.61	0.12	-1.00	-0.25
Artificial_hi	4	6.4	0.8	3.9	9.1	2.08	0.19	1.10	2.47
Artificial_li	4	8.9	0.8	6.0	11.3	2.01	0.22	1.39	2.82
Forest_hi	2	7.0	1.1	3.7	10.8	0.78	0.15	0.34	1.28
Forest_li	1	33.4	5.3	15.6	53.0	-0.36	0.11	-0.71	-0.06
Non-use	1	19.8	2.4	10.9	27.0	-0.45	0.27	-1.47	0.34
Linear model – threatened species									
Agri_hi	0	–	–	–	–	–	–	–	–
Agri_li	2	72.6	3.7	61.3	83.5	-1.40	0.32	-2.44	-0.36
Artificial_hi	4	37.1	2.6	27.9	44.8	4.18	0.58	2.47	6.08
Artificial_li	4	42.2	2.1	35.5	49.4	3.59	0.58	1.47	5.45
Forest_hi	1	30.8	5.9	10.3	49.0	2.61	0.79	0.52	5.28
Forest_li	1	73.7	10.3	37.0	107.8	-0.54	0.23	-1.30	0.16
Non-use	0	–	–	–	–	–	–	–	–
Nonlinear model – all species									
Agri_hi	2	1.8	0.1	1.3	2.3	0.03	0.01	0.01	0.05
Agri_li	2	3.1	0.1	2.7	3.6	-0.08	0.01	-0.11	-0.04
Artificial_hi	4	1.8	0.1	1.5	2.2	0.18	0.02	0.08	0.26
Artificial_li	4	2.0	0.1	1.7	2.2	0.17	0.02	0.11	0.24
Forest_hi	2	1.8	0.1	1.3	2.1	0.08	0.02	0.04	0.15
Forest_li	1	3.9	0.5	2.1	5.2	-0.03	0.01	-0.06	0.00
Non-use	1	2.9	0.2	2.3	3.6	-0.04	0.02	-0.10	0.02
Nonlinear model – threatened species									
Agri_hi	0	–	–	–	–	–	–	–	–
Agri_li	2	4.4	0.1	4.0	4.7	-0.04	0.01	-0.06	-0.02
Artificial_hi	4	3.5	0.1	3.3	3.8	0.10	0.02	0.06	0.15
Artificial_li	4	3.6	0.1	3.5	3.8	0.08	0.02	0.04	0.13

		Parameter *a*				Parameter *b*			
	n	Mean	Std.Dev	Min	Max	Mean	Std.Dev	Min	Max
Forest_hi	1	3.2	0.2	2.5	3.9	0.07	0.02	0.01	0.14
Forest_li	1	4.4	0.3	3.7	5.3	-0.01	0.01	-0.05	0.03
Non-use	0	–	–	–	–	–	–	–	–

Tab. 7-15: Regional damage/benefits calculated for ***all*** *species according to equations [21] and [22].*

	linear damage function	nonlinear damage function					
Types of intensity	$EDP_{region.-lin}$ (= *b* mean)	*a* mean	*b* mean	*L* [%] plateau	*L* [%] c-alps	$EDP_{region.-nonl-pla-4}$	$EDP_{region.-nonl-c-alp}$
Agri_hi	0.38	1.80	0.03	49	7	0.79	0.23
Agri_li	-0.61	3.13	-0.07	9	20	-0.86	-0.39
Artificial_hi	2.08	1.76	0.18	8	2	4.04	1.55
Artificial_li	2.01	1.97	0.17	7	1	3.68	1.47
Forest_hi	0.79	1.76	0.08	9	3	0.97	0.60
Forest_li	-0.36	3.85	-0.03	18	46	-0.83	-0.36
Non-use	-0.46	2.94	-0.04	0	18	-0.74	-0.36

where a mean, b mean: mean of parameter used in the damage function (equations [21] and [22]); $EDP_{region.-lin}$: regional damage potential with linear model; $EDP_{region.-nonl}$: regional damage potential with nonlinear model; pla-4: part of Swiss Plateau; c-alp: Central Alps; L[%]: actual proportion of land use type in %.

Tab. 7-16: Regional damage/benefits calculated for ***threatened*** *species according to equations [21] and [22].*

	linear damage function	nonlinear damage function					
Types of intensity	$EDP_{region.-lin}$ (= *b* mean)	*a* mean	*b* mean	*L* [%] plateau	*L* [%] c-alps	$EDP_{region.-nonl-pla-4}$	$EDP_{region.-nonl-c-alp}$
Agri_hi	–	–	–	49	7	–	–
Agri_li	-1.42	4.39	-0.04	9	20	-2.27	-1.43
Artificial_hi	4.16	3.50	0.10	8	2	7.01	4.13
Artificial_li	3.58	3.63	0.09	7	1	5.57	3.52
Forest_hi	2.61	3.24	0.07	9	3	3.40	2.22
Forest_li	-0.54	4.42	-0.01	18	46	-0.70	-0.53
Non-use	–	–	–	0	18	–	–

7.7 Total characterisation factor (EDP_{total})

7.7.1 Combination of local and regional damage/benefits

In order to calculate the total aggregated damage for a land use activity it is necessary to combine the local and the regional damage/benefits. For doing this two options are possible. Option 1 is to regard *EDP* as a dimensionless factor, which is used to weight a land use activity with respect to its ecological consequences. Option 2 is to calculate the species potentially lost as a result of a land use activity using the characterisation factor *EDP*. The main difference between these is that aggregation of local damage and regional damage is either based on subjective weighting (Option 1) or on an analytical approach (Option 2). A numerical example for both options is given in chapter 8.1 (p. 139).

Option 1

To get a single index two or more indicators are added together or multiplied. When the index is arrived by multiplication, the total ecosystem damage potential (EDP_{total}) is calculated as

$$EDP_{total} = k * EDP_{local} * EDP_{regional} \qquad [23]$$

Certain choices made before render this option impossible. The main reason is that both EDP_{local} and $EDP_{regional}$ can adopt negative values, which indicates a benefit in both cases. Multiplying two negative values (benefits) would result in a positive value (damage), which does not make sense.

When the index is arrived at by addition, the index formation is named a weighted sum method. It includes the performance indicators (*EDP*), values to standardise those performance indicators (n), and weighting factors (k) expressing the relative importance of one unit of the standardised performance indicators. It is necessary to standardise the values for EDP_{local} and $EDP_{regional}$ to values between -1 and 1, because this eliminates effects of unit changes. After standardisation a sound weighting of both indicators is possible.

$$EDP_{total} = k_l \frac{EDP_{local}}{n_l} + k_r \frac{EDP_{regional}}{n_r} \qquad [24]$$

The standardisation is done with n_l and n_r, the weighting of the local and the regional *EDP* with k_l and k_r.

The default weighting of EDP_{local} and $EDP_{regional}$ is 50% each. However, deviation is possible if assessors want to weight the two basic values of biodiversity (chapter 7.2, pp. 62) differently. If one focuses, for example, on the conservation value of biodiversity, the regional EDP can be weighted more compared to the local one.

I calculated two types of total *EDP*: a linear one using option 1 for EDP_{local} and $EDP_{regional}$, and a nonlinear one using option 2 for both EDPs. The linear and nonlinear option for EDP_{local} are described in the chapter 7.5.3. Those options for $EDP_{regional}$, are described in chapter

7.6.4. Calculating two types of total EDP provides the opportunity to assess differences in the result.

Option 2

The second option stresses the possibility to calculate the absolute number of species lost due to a land use activity for the local as well for the regional scale. On the local scale the *EDP* factor equals the proportion of species missing on a land use type *i* in comparison to a reference. Thus the factor can be used to calculate the number absolute number of species missing for a specific land use type. On the regional scale the *EDP* factor equals the potential loss of species in % when a specific type of intensity of land use increases by 1%. Also in this case it is possible to calculate the number of species potentially lost. Assuming that the damage "10 species lost on 1 km^2" is equivalent to the damage "1 species lost on 10 km^2" it is possible to add together the local and regional damage without any weighting.

For one or the other option different arguments apply. One could argue that the reason for considering both local and regional impacts is that the range of values involved is wider. In this case local and regional damage are different as a matter of principle and weighting of these two is necessary (option 1). Another line of reasoning is that land use triggers damage on two scales. One is a local scale and one a regional scale but, in principle, the damage is the same. Because of this, it is possible to aggregate these types of damage based on an analytical procedure (option 2). Whether option 1 or option 2 is appropriate can only be decided by looking at it in terms of consistency. I decided to develop option 1 further since this allows focusing on either EDP_{local} or $EDP_{regional}$, by putting different weights on them.

7.7.2 Results for EDP_{total}

The total damage potential represents the characterisation factor used in the impact assessment of land use activities. It was constructed in an additive mode according to option 1 (equation [24], p. 127). The local ecosystem damage potential EDP_{local} could be used without standardisation. The regional damage potential for all species $EDP_{regional}$ was divided by n_r = 4.41 to get the standardised *EDP**, which is in the range from -1 to +1. The regional damage potential for threatened species was divided also by n_r = 4.41. This has the consequence that it is possible to compare both the EDP_{total} for all species and the EDP_{total} for threatened species (see appendix 6a). When focusing on one land use type the values in appendix 6a show that values of *EDP** for threatened species are more extreme than those for all species. *EDP** calculated for threatened species are more negative (=higher benefit) for beneficial land use types and more positive (=higher damage) for damaging land use types compared to the same *EDP** calculated for all species. In the case of damaging land use types the *EDP** for threatened species can even be above +1 since this is the maximum value which was set for *EPD** of all species. In every case of calculating the EDP_{total} the local and the regional part of the total damage potential were considered with a weight of 50% (cf., equation [24]: $k_l = k_r = 0.5$) each.

Different versions of the total characterisation factors EDP_{total} were computed for comparative reasons (Fig. 7-21, p. 131). It is possible to distinguish the kind of principal modelling approach (linear versus nonlinear), and the reference value (Swiss Plateau versus Central Alps). The detailed results for the different species groups (all species versus threatened species) can be found in appendix 6 (p. 237).

Focusing on EDP_{local} one can see in Fig. 7-21a that factors calculated with the linear model (lin) cover a wider range of values compared to those calculated with the nonlinear (nonl) model. This is because the effect-damage curves in Fig. 7-8 (p. 92), which were applied, differ in their shape. The nonlinear effect-damage function transforms an equal delta of two land use type's relative species number into a different delta of corresponding damage potentials depending on the level of species richness. In principle, a delta between two species rich land use types becomes a larger delta EDP_{local} compared to a delta between two species poor types. The rationale behind this is that the loss of a number of species on a given plot which is anyway poor in species triggers a large damage.

Generally speaking, the characterisation factors EDP_{local} calculated for the Plateau (pla) are smaller compared to the factors EDP_{local} for the Central Alps (c-alp). This is because the regional species number, which was used to get a relative species number, was higher for the Central Alps than for the Plateau. However, I did not differentiate the land use type's species numbers with respect to the regions rather then taking the same values for all regions.

Focusing now on the $EDP_{regional}$ (Fig. 7-21b) it is striking that the differences for artificial areas between different $EDP_{regional}$, which are based on different models, are rather large. This is because the slope parameter b used to calculate the $EDP_{regional}$ with the nonlinear model is rather large for the artificial surfaces (Tab. 7-15 p. 125, see also effect-damage curve in Fig. 7-20b). This means that the differences in the proportion of artificial areas between the regions results in large differences in corresponding $EDPs_{regional}$.

Total characterisation factors EDP_{total} are calculated as the arithmetic average between EDP_{local} and $EDP_{regional}$. All types of EDP_{total} tend to be more equal (Fig. 7-21c) than expected. Previous differences between types of *EDPs* both on the local and the regional scale are levelled. At the same time the factors EDP_{local} and $EDP_{regional}$ complement one another resulting in a sound ranking of land use types.

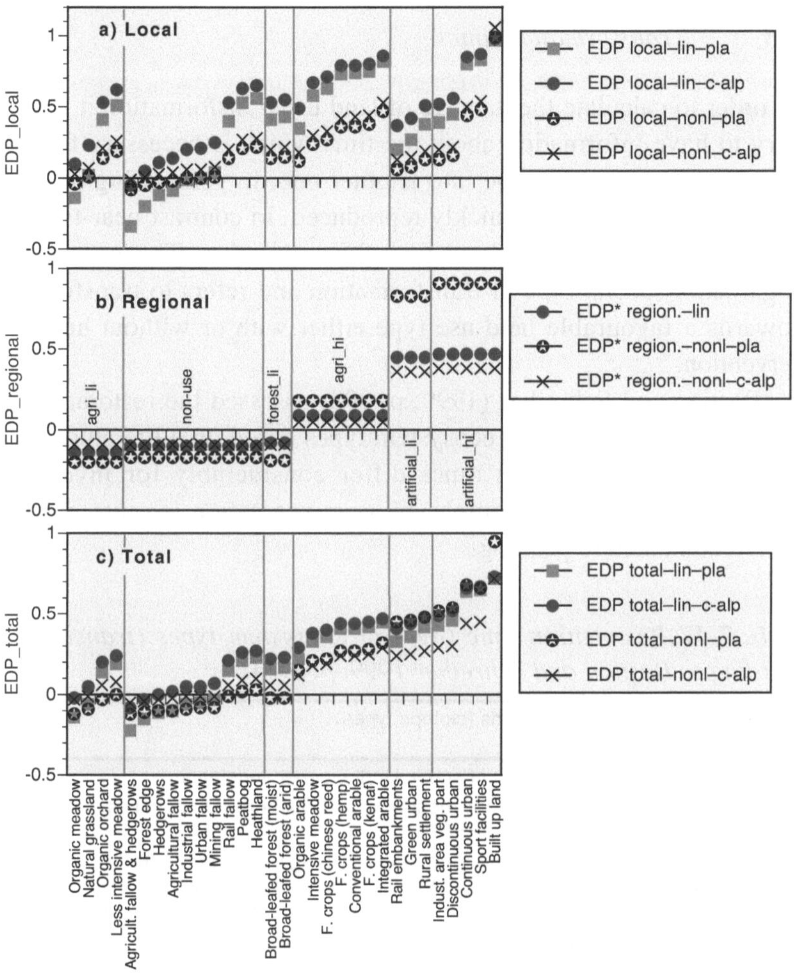

Fig. 7-21: Local, regional, and total ecosystem damage potential (EDP) for all species. For the regional scale, I used the standardised EDP which equals 1/4.41*EDP. Local and regional damage factors are each weighted with 50%. (exact figures are given in appendix 6a, p. 237)*

7.8 Transformation times

In order to calculate the damage of land use transformation it is necessary to have information about the time which is necessary for transforming one land use type into another one. In general high-intensity land use types are very quickly reproduced. In contrast near-to-natural land cover types need a long time for restoration. The restoration is regarded a special type of transformation and refers to transformation towards a favourable land use type either with or without human intervention.

Bastian and Schreiber (1999, p. 298) assessed the restoration time for a number of different ecosystem types based on a literature review (Tab. 7-17). Restoration times differ considerably for investigated ecosystem types. For example, pioneer vegetation is quite easily restored compared to peatbog.

Tab. 7-17 Restoration time (age) of ecosystem types (translated according to Bastian and Schreiber 1999, p. 298).

Restoration time (years)	Ecosystems (biotope types)
< 5	Vegetation of arable land, pioneer vegetation
5 – 25	Species poor meadows and tall-herb communities, mature pioneer vegetation
25 – 50	Species poor immature hedgerows and shrubs, oligotroph vegetation of areas silting up, relatively species rich marshland with sedges, meadows, dry meadows and heathland
50 – 200	Forests quite rich in species, shrubs and hedgerows
200 1'000	Low and medium (immature) peatbogs, old dry meadows and heathland
1'000 – 10'000	High (mature) peatbogs, old growth forests

Based on the data given in Tab. 7-17 I estimated the restoration time of different types of intensities of land use (Tab. 7-18). It should be clear that the estimates are somewhat arbitrary and only a rough first approach. Depending on the initial/final quality of the land use type, the climate, and many other factors, restoration times may vary quite considerably. Some ecosystem types also might even never be restored again.

Tab. 7-18: Estimated time in years which is necessary to transform an initial land intensity into a final land intensity.

	Final →						
Initial ↓	Agri_hi	Agri_li	Artificial_hi	Artificial_li	Forest_hi	Forest_li	Non-use
Agri_hi	–	10	<1	2	25	50	500
Agri_li	<1	–	<1	2	25	50	500
Artificial_hi	5	10	–	2	25	50	500
Artificial_li	2	5	<1	–	25	50	500
Forest_hi	2	2	<1	2	–	25	?
Forest_li	2	2	<1	2	10	–	?
Non-use	<1	<1	<1	2	10	25	–

Agri_hi: conventional arable, integrated arable, organic arable, fibre/energy crops, intensive meadow;

Agri_li: less intensive meadow, organic meadow, organic orchard, natural grassland

Artificial_hi: built up land, continuous urban, discontinuous urban, sport facilities, industrial area - part with vegetation;

Artificial_li: green urban, rural settlement, rail embankments;

Forest_hi: forest plantations;

Forest_li: semi-natural broad-leafed forest (either moist or arid);

Non-use: heathland, hedgerows, peatbog.

7.9 Data uncertainties

7.9.1 Monte-Carlo simulation to model data uncertainties

For modelling data uncertainties of the final results, which are the characterisation factors (*EDP*), a Monte-Carlo simulation was applied. This numerical method is based on random sampling and allows calculating a final distribution of output variables based on the distribution of input variables (Morgan and Henrion 1990; Sobol 1994). To run the simulation the software package @Risk (Palisade Corporation 1997) was used. Important input variables for each land use type were the relative species number on the local scale and the percentage of species potentially lost on the regional scale. Output variables were characterisation factors (ecosystem damage potential *EDP*), which are specific for different land use types. In total, each simulation had 1'000 iterations, which means that the final distribution of the output variables is based on 1'000 data points. The sampling applied was of the continuous Monte-Carlo type and not the stratified Latin Hypercube type.

7.9.2 Results of the Monte-Carlo simulation

The Monte-Carlo simulation results in the mathematical distribution of the characterisation factor *EDP* (ecosystem damage potential). Its distribution is dependent on the distribution of all input variables of the model. The result show for the linear model taking all the species into account that in general the standard deviation increases with decreasing *EDP* (Fig. 7-22a and b). A possible explanation for this is that land use types that are in general high in ecosystem quality show a rather high variability of species diversity. In contrast ecosystem types with low ecosystem quality show a low but constant species diversity.

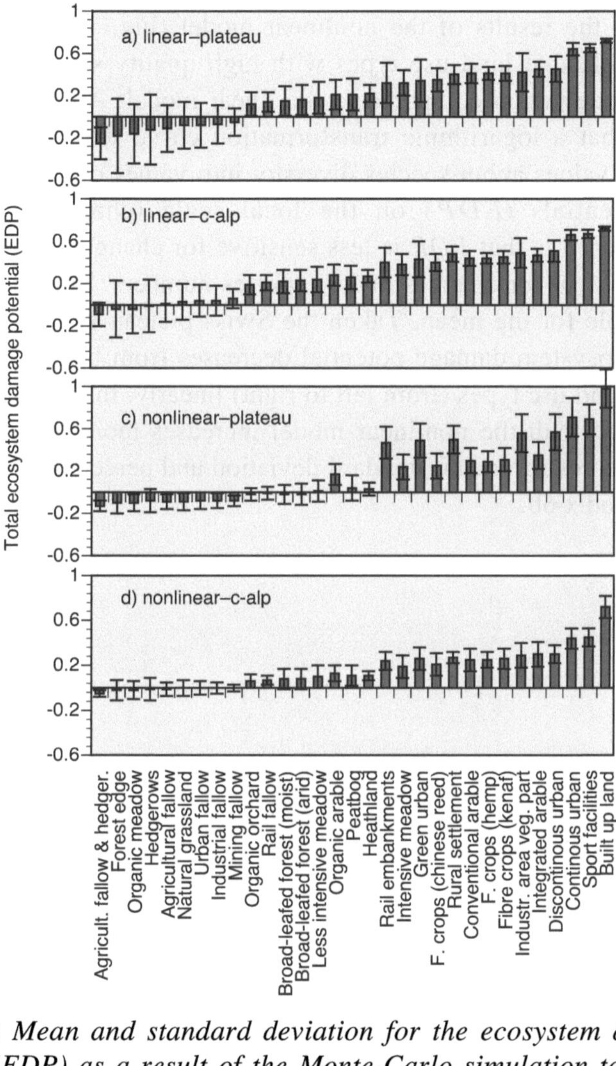

Fig. 7-22: Mean and standard deviation for the ecosystem damage potential (EDP) as a result of the Monte-Carlo simulation taken all species into account. I distinguished in a) to d) the linear/nonlinear model and the Swiss regions (plateau/c-alps).

Looking the results of the nonlinear model (Fig. 7-22c and d) it is remarkable that land use types with high quality show a rather low standard deviation compared to the linear model. An explanation for this is that a logarithmic transformation curve was used for transforming values about species diversity into values of ecosystem damage potentials (*EDP*) on the local scale (chapter 7.5.3). The consequence is that *EDP* is less sensitive for changes of high species numbers than for changes of low species numbers. This effect is also observable for the mean. Taken the Swiss plateau as an example the mean ecosystem damage potential decreases from beneficial to detrimental land use types (from left to right) linearly. In contrast the mean calculated with the nonlinear model increases more logarithmically. Exact figures for mean, standard deviation and percentiles are given in the appendix 6b.

Part C

Application of proposed land use assessment method

8 Calculating the final damage with exemplary inventory data

In this chapter two examples are provided which show how characterisation factors are applied in the calculation of the total damage resulting from a specific land use activity. The first example deals with the occupation of a land use type. The second example shows how to calculate the damage from a more complex land use activity, one that has a transformation phase, a occupation phase and a restoration phase.

8.1 Occupation of conventional arable land

In the first example, the damage due to occupation of conventionally managed arable land is calculated. The information given in the sample Life Cycle Inventory is that 20 hectare of arable fields are occupied for 1 year to produce grain in the Swiss Plateau. The product of area and time (surfacetime) is then weighted with the characterisation factors, which refer to the ecosystem damage potential with a focus on species diversity (*EDP*). For comparative reasons the both options for applying *EDP* described in chapter 7.7.1 (pp. 126) are elaborated.

Option 1

Generally, the damage is calculated according to equation [4]. It is the product of the characterisation factor, the area of occupation and the duration of occupation. Local characterisation factors (EDP_{local}) for conventionally managed arable land are documented in Tab. 7-7 (p. 95) and regional ones ($EDP_{regional}$) in Tab. 7-15 (p. 125). To calculate the total damage, the standardised $EDP_{regional}$* is used, that is the $EDP_{regional}$ divided by the maximum $EDP_{regional}$ (about 4.41). To calculate the total damage, local and regional damage are weighted at 50% each and then added together.

The total damage D_{total}, which shows the unit [ha*yr*EDP], is calculated as follows:

	Local			Regional		
	Surface-time [ha*yr] (1)	$EDP_{local\_lin\_pla}$	D_{local} (2)	$EDP_{regional\_lin\_pla}$* (3)	$D_{regional}$ (4)	D_{total} (5)
Conventional arable/Agri_hi	20	0.74	14.8	0.09	1.8	8.3

(1) The product of area of occupation (20 ha) and time (1 yr).

(2) Local damage to EQ due to occupation: 0.74*20 ha*1 yr

(3) $EDP_{regional\_lin\_pla}{}^{*} = EDP_{regional\_lin\_pla}/\mathrm{Max}(EDP_{regional\_lin\_pla}) = 0.38/4.41$

(4) $D_{regional}$ = 0.09*20 ha*1 yr

(5) $D_{total} = 0.5\ D_{local} + 0.5\ D_{regional}$

Option 2

For this option the characterisation factors are used to calculate the species potentially lost for both the regional and the local scale. The total damage has the unit [species lost*ha*yr] and was calculated as follows:

	Local			Regional			
	Surface-time [ha*yr] (1)	Local S_{lost} (2)	D_{local} (3)	Surface-time [ha*yr] (4)	Regional S_{lost} (5)	$D_{regional}$ (6)	D_{total} (7)
Conventional arable/Agri_hi	20	30	600	10000	0.0104	104	704

(1) The product of the area occupied (20 ha) and the duration of occupation (1 yr).

(2) The local number of species lost (S_{lost}) is the $EDP_{local\_lin\_pla}$ (0.75) multiplied by the reference number of species (40). Here, what is ment by $EDP_{local\_lin\_pla}$ is that compared to the reference 75% of the species on conventional arable land are missing.

(3) Local damage D_{local} is the local surfacetime multiplied by the number of species lost.

(4) The regional surfacetime is the size of the polygon (about 100 km^2 in the Swiss Plateau) multiplied by the time of usage (1 yr). The size of the polygon is taken because the $EDP_{regional\_lin\_pla}$ refers to this affect area.

(5) According to the $EDP_{regional\_lin\_pla}$ an increase in the area of high-intensity agriculture of 1% results in a potential loss of 0.38% of species on the regional scale. On average, each polygon of the Swiss Plateau has 679 species (Tab. 7-5, p. 86) indicating

that about 2.6 species are lost. However, only 20 ha are occupied, which is equal to 1/5, or 20% of 1% of the polygon area. Therefore, only 1/5 of 2.6 can be assigned to the land use activity, which should be assessed. Because I assume that it takes 50 years for the species loss to manifest itself, only 1/50 of the percentage of species lost can be assigned to the 1 year of occupation. Therefore, the regional number of species lost due to the land use activity is 1/5*1/50*2.6=0.0104.

(6) The regional damage is calculated as the product of the surfacetime (4) and the regional number of species lost due to the land use activity (5).

(7) The total damage equals the sum of the local and regional damage $D_{total} = D_{local} + D_{regional}$

Though the approaches of option 1 and option 2 are different, their numerical results are quite similar. In both cases, the local damage is greater than the regional damage. Even the proportion of one type of damage to the other is similar. Applying option 1 about 12% of the total damage is assigned to the regional damage and applying option 2 it is about 17%. Although option 2 employs a more analytical approach, option 1 has the advantage of easier application. The characterisation factors which can be used for option 1 are given in appendix 6.

8.2 Building and using a commercial plant in a forest

In the second example, I focus on a more complex sequence of land use activities, including land transformation. The information in the sample Life Cycle Inventory is that a low-intensity forest of 150 ha is transformed into a high-intensity artificial area, which is used for another 7 years. It is assumed that land use takes place in a region with land use proportions comparable to the Swiss Plateau (that is, rather intensive land use). Since transformation is involved in this example the delta change of damage potential in time (refers to improvements/deterioration in ecosystem quality) and the damage integrated over time can be distinguished (chapter 7.4, pp. 66) . The integrated damage is divided into the damage from transformation, the damage from occupation and the damage from a virtual restoration phase (cf.,

Fig. 7-3a). The damage from occupation was calculated according to equation [4] and that from transformation according to equation [5]. To calculate the damage from the restoration phase equation [5] and the data on restoration times were used (Tab. 7-18).

The total integrated damage was computed as follows:

				Local		Regional		
	Area [ha]	Time [years] (1)	Surface-time (2)	$EDP_{local\_lin\_pla}$ (3)	D_{local} (4)	$EDP^*_{regional\_lin\_pla}$ (5)	$D_{regional}$ (6)	D_{total} (7)
Transformation								
from Forest_li to Artificial_hi	150	0.5	75	0.70	53	0.20	15	
Occupation								
Artificial_hi	150	7	1050	0.97	1019	0.47	494	
Restoration								
from Artificial_hi to Forest_li	150	50	7500	0.70	5250	0.20	1463	
Total			8625		6321		1971	4146

(1) The duration of transformation is 0.5 years and that of virtual restoration is 50 years.

(2) The surfacetime is the product of area occupied and duration of occupation [ha*yr].

(3) Local characterisation factors for the plateau, taking the linear model into account ($EDP_{local\_lin\_pla}$) are given in Tab. 7-7 (p. 95) and appendix 6 (pp. 237). I used the more specific local factors for built up land (*EDP*=0.97) and semi-natural broad-leafed forest (*EDP*=0.43). Factors for transformation and restoration are calculated as the mean of factors of initial and final land use types.

(4) The local damage equals the product of the surfacetime (2) and local characterisation factor (3).

(5) The standardised regional factors $EDP^*_{regional\_lin\_pla}$ can be found in appendix 6. The *EDP* for artificial_hi is 0.47 and this for forest_li -0.08. Factors for land use changes are calculated according to (3).

(6) The regional damage $D_{regional}$ is the product of the surfacetime (2) and regional characterisation factor (5).

(7) The total damage is the sum of local and regional damage D_{total} = 0.5 D_{local} + 0.5 $D_{regional}$.

The results show that a lot of damage results from the local scale restoration (5250 points). This is because the time necessary to increase ecosystem quality in terms of species diversity to the initial level is rather long time (50 years for restoration phase). In addition, the built-up land I took into account on the local scale scores worse than all other land use types. This virtual restoration part of the damage is an immediate consequence of the transformation and is necessarily to be included when land use transformation is assessed.

The delta change of damage potential in time can be calculated as well. It was defined as the difference in the damage potential at the beginning and the end of the land use activity (equation [6], p. 68). Since in this example it was assumed that the initial land use type and the final land use type are both low-intensity forest the net difference in the ecosystem damage potential is zero.

9 Case study: Land use impact assessment of food products

The purpose of this small case study about food products is to show how the method proposed for assessing land use works in an application. It also clarifies which land use data are necessary for a Life Cycle Inventory. However, this case study does not provide an in-depth Life Cycle Impact Assessment (LCIA) of food products, because only land use was taken into account. Other types of environmental interventions, like energy input or chemical emissions, are not considered at all in this example.

9.1 Land use in the life cycle of food products

Food products, like meat or vegetables, are produced in the agricultural sector. This sector requires large areas of a country's surface to be devoted to production of biomass, a major input to food products. Although agricultural land use is very important in the life cycle of food products, these products' life cycle also includes phases like transportation and processing, which involve non-agricultural land.

The vegetables' life cycle includes phases in which fresh vegetables are *produced*, *stored* by retailers, *transported* from Europe to Switzerland, and *packaged* in plastic foil. In the case study, three modes of production are compared: in a greenhouse, in a less intensive mode outdoors, and organically outdoors. Transportation, storage, and packaging are assumed to be the same for all alternatives. The formulation of the alternatives is based on the method of modular LCA by Jungbluth (2000).

There are two categories of land use as well as use of marine benthos involved:

- Direct land use for agricultural production, and transport via road, railway or airport.
- Indirect (grey) land use, e.g., for production of fertiliser, refinement of petrol, and extraction of coal.
- Use of marine benthos (sea bottom) for extraction of crude oil and transport via pipelines in the sea.

9.2 Inventory data

The assessment was made using 1 kg vegetables as the functional unit. A mix of vegetable consisting of beans, leek, celery and tomatoes was taken into account. In the inventory the land use is registered which is necessary for production, storage, packing, and transportation of this amount of vegetable. The land use is measured in terms of surface and time or, more precisely, as their product, the sufacetime with units of [m^2*yr]. All inventory data is taken from Jungbluth (2000). The database *Ecoinvent*[12] has been used by Jungbluth as a reference for background data (Frischknecht et al. 1996) and he completed this database with data on agricultural land use.

In the inventory, land occupation and transformation are taken into account. Land occupation is due to agricultural land use and due to occupation with infrastructure (*Ecoinvent type IV*). In these cases, it is

[12] Definitions of land use types in the database Ecoinvent are:

- Ecoinvent type I: natural land, impact of human beings not larger than those of other species
- Ecoinvent type II: modified land, impact of human beings larger than those of other species, but non-cultivated components (e.g., near-to-nature forests)
- Ecoinvent type III: cultivated land, impact of human beings larger than those of other species, mostly cultivated components (e.g., agriculture, forest plantations)
- Ecoinvent type IV: built up land, dominated by streets, buildings, industrial plants.

not assumed that restoration is actual. Mining operations, however, are a typical case of land transformation, because the exploited mine is actually restored. Mining operations include these phases: construction, exploitation, destruction and restoration (Fig. 9-1). In the *Ecoinvent* database two cases of transformation are distinguished for mining operations.

- Mining operations starting with *Ecoinvent type II* as their initial land use type. Here the surfacetime of the construction, exploitation, destruction and restoration which corresponds to a change from *Ecoinvent type II* into *Ecoinvent type IV* and back into *Ecoinvent type III* again, is registered as *Ecoinvent type II->IV*. The restoration from *Ecoinvent type III* back to *Ecoinvent type II* are registered as *Ecoinvent type II->III*.
- Mining operations showing the worse land use type *Ecoinvent type III* at the beginning. In this case, the transformation from *Ecoinvent type III* to *Ecoinvent type IV* back to *Ecoinvent type III* again – involving stages of construction, exploitation, destruction, and restoration – are registered *as Ecoinvent type III->IV*.

In the *Ecoinvent*, it is assumed that 5 years are necessary for *Ecoinvent type IV* to be restored to *Ecoinvent type III*, 50 years for *Ecoinvent type III* to be restored to *Ecoinvent type II,* and 100'000 years for *Ecoinvent type II* to be restored to *Ecoinvent type I*. However, restoration into *Ecoinvent type I*, which is natural land, is not considered in the *Ecoinvent*.

In the inventory available for this assessment only the surfacetime (including the restoration time) is registered, but it was not possible to split this up into the surface and the time.

The results of the inventory are shown in Fig. 9-2 in the graphs in the left column (exact figures are given in appendix 7). The differences between greenhouse and outdoor production (whether organic or less intensive) are quite remarkable. In the case of outdoor production about 90% of the surfacetime is agricultural land use. In contrast, in greenhouse production only 32% of the total surfacetime can be assigned to use of the greenhouse, 36% can be assigned to the use of marine benthos for extraction of crude oil, and 18% can be assigned to transportation. Storage and packaging are not important for any alternative.

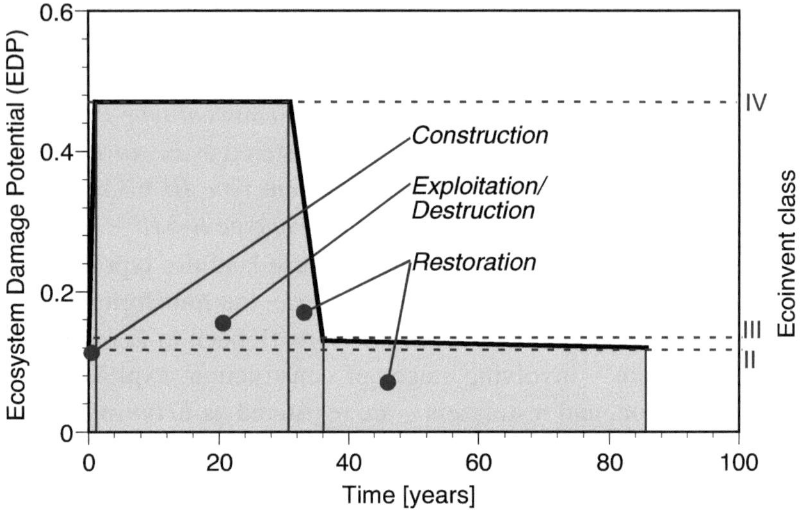

Fig. 9-1: A mining operation as an example of land transformation with different phases of land use. Characterisation factors EDP are taken from appendix 6a.

9.3 *Impact Assessment using characterisation factor EDP*

In general, characterisation factors are used in an impact assessment to take the differences in ecological impact of different land use types into account. The factors used in the case study are derived from the simple linear model of the Swiss Plateau, where local and regional damage are included (Tab. 9-1). This is the most basic version of the total characterisation factor. More information about the selection of EDPs can be found in chapter 13.2. Factors for land transformation are calculated according to equation [5].

The impact assessment results in changes to the ranking of alternatives (Fig. 9-2, graphs in right column). The alternatives are: vegetables produced in greenhouses (*greenhouse vegetables*), vegetables produced less intensively outdoors (*less intensively grown vegetables*), and vegetables produced organically outdoors (*organic vegetables*). When alternatives are ranked on the basis of their pure surfacetime [m^2*yr], then *greenhouse vegetables* perform best. *Less intensively grown vegetables* and *organic vegetables* are quite equal to one another, but worse than *greenhouse vegetables*. However, the picture changes when the characterisation factors for weighting the inventory data are applied. Here *greenhouse vegetables* and *organic vegetables* are equal. However, their modes of production are almost opposite. In the first case, the area is very small, but its ecosystem damage potential is quite high, because production is very intensive. In the second case, the ecosystem damage potential is less, but a larger area is necessary to produce the same amount of vegetables. The land area which is used as organic meadow is even be regarded as a benefit for ecosystem quality and shows a negative score. *Less intensively grown vegetables* are much worse than others, because of the rather high ecosystem damage potential of less intensive agriculture.

Tab. 9-1: Characterisation factors ($EDP_{total\_lin\_pla}$) from appendix 6a.

Land use activity	$EDP_{total\_lin\_pla}$
Land occupation	
Arable land, less intensive	0.45
Arable land, organic	0.22
Meadow, less intensive	0.19
Meadow, organic	-0.14
Built up land	0.72
Ecoinvent type II	0.12
Ecoinvent type III	0.13
Ecoinvent type IV	0.47
Land transformation	
Ecoinvent type II-III: (II+III)/2	0.125
Ecoinvent type II-IV: (II+IV)/2	0.295
Ecoinvent type III-IV: (III+IV)/2	0.3

Greenhouse production requires a lot of heating with oil fuel. The extraction of crude oil uses about the same size surface as the area for the construction of the greenhouse. However, transformation of marine benthos is not considered in the impact assessment, since no characterisation factors are available. If marine benthos were considered, *greenhouse vegetables* would perform worse than *organic vegetables*.

A major problem was that in *Ecoinvent* the classification into land use types is based on their degree of naturalness. For example, *Ecoinvent type II* and *Ecoinvent type III* show different degrees of naturalness, but both are very similar in terms of EDP. This is because their total number of species is similar. Therefore, the usefulness of making links between *Ecoinvent* inventory data and the characterisation factor EDP is limited. For future inventories, it would be advisable that the data from forestry, agriculture, and artificial land is at least based on land intensity (high- or low-intensity, cf., Appendix 4).

Please note that the ranking does not apply to every vegetable but to the mix of vegetable consisting of beans, leek, celery and tomatoes. Ranking of alternatives of production for other vegetables than these can vary considerably.

*Fig. 9-2: Life Cycle Assessment of 1 kg of vegetables produced in greenhouses, outdoors in less intensive mode and outdoors in organic mode. The left column of charts shows the surfacetime [m^2*yr] registered in the inventory and the right column of charts shows the weighted surfacetime using the factor EDP (see next page).*

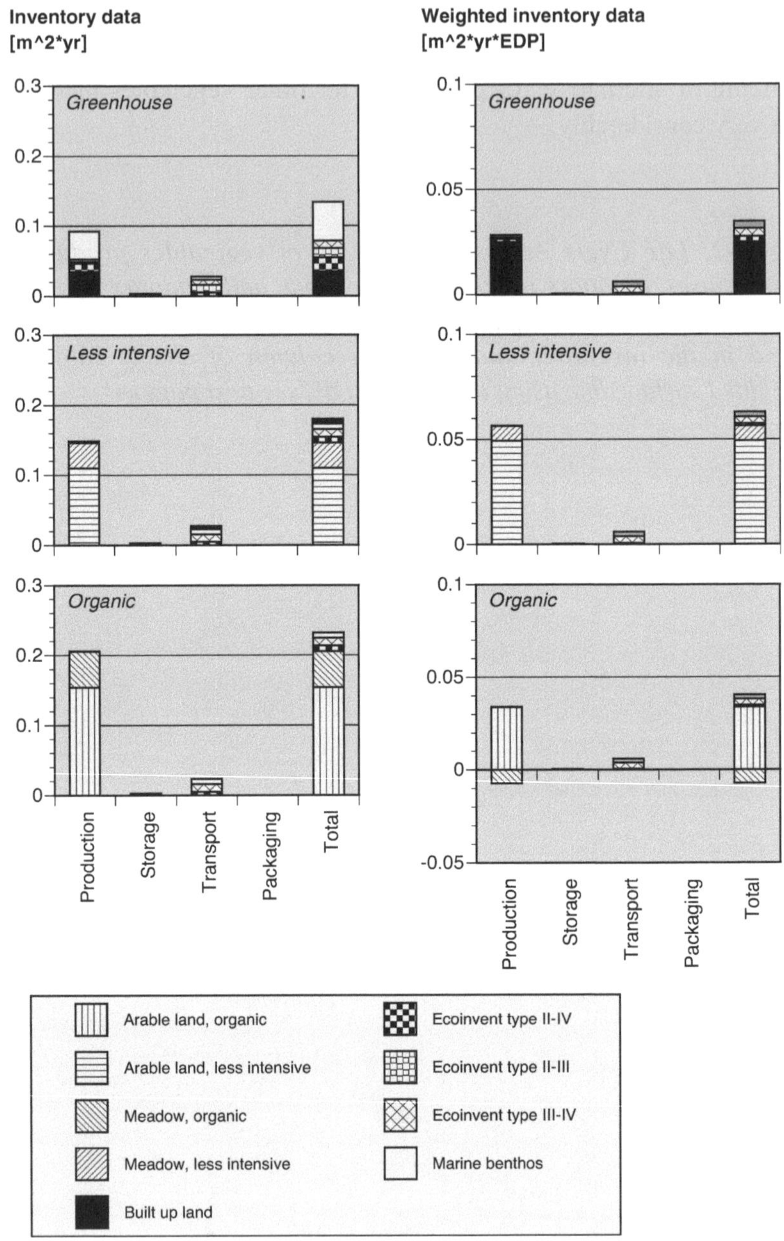
Inventory data
[m^2*yr]
Weighted inventory data
[m^2*yr*EDP]
Greenhouse
Less intensive
Organic
0.3
0.2
0.1
0
0.1
0.05
0
-0.05
Production
Storage
Transport
Packaging
Total
Arable land, organic
Arable land, less intensive
Meadow, organic
Meadow, less intensive
Built up land
Ecoinvent type II-IV
Ecoinvent type II-III
Ecoinvent type III-IV
Marine benthos

10 How to integrate one's data on species richness of different land use types

A person can use his own empirical data about species numbers per land use type to calculate characterisation factors for the local scale. For example, if one wants to consider the species number of forests, forest edges, and agricultural land, then he needs to know their species numbers per area, which is about 20 species/120 m^2 in forests, 42 species/50 m^2 at forest edges, and 24 species/32 m^2 on agricultural land (numbers are taken from Flückiger 1999). Though one finds almost the same species numbers in forest and agricultural land, they cannot be compared, because they refer to quite different plot sizes.

1) Standardised species number per area in ha

In order to take the differences in the sizes of areas into account, it is necessary to calculate the standardised species number per standard area (0.01 ha = 100 m^2). Due to the species-area relationship, one expects to find on 100 m^2 more than 24 species on agricultural land (where the area investigated was only 32 m^2 = 0.0032 ha) and less than 20 species in forests (where the area was 120 m^2 = 0.012 ha). According to equations [10] and [11] the changes in species number are calculated while changing the plot size. In order to calculate the standardised species number for agricultural land it is first necessary to calculate the average species number S on 32 m^2 of all land use types as follows:

- $\ln S_{average}(A) = 4.1+0.2*\ln A$ (chapters pp. 71 and pp. 91)
- with $A = 0.0032$ ha
- $\ln S_{average}(0.0032 \text{ ha}) = 4.1+0.2*\ln 0.0032$ ha
- $\ln S_{average}(0.0032 \text{ ha}) = 2.95$ ha
- $S_{average}(0.0032 \text{ ha}) = 19$

For the standardisation area average S is calculated as above:

- $S_{average}$ (0.01 ha) = 24

Given any land use type, one finds an average of 19 species per 32 m^2 and 24 species per 100 m^2; that is, there is a difference of 5 species. I made the assumption that this relationship applies to any given specific land use type. The standardised species number for agricultural land is calculated as the sum of the number of species found on 32 m^2 and the expected increase in species from sampling 100 m^2 rather than 32 m^2. This is:

- $S_{agricultural\ land}$ (0.01) = $S_{agricultural\ land}$ (0.0032 ha) + [($S_{average}$ (0.01 ha) - $S_{average}$ (0.0032 ha)]
- $S_{agricultural\ land}$ (0.01) = 24 + (24-19) = 24 + 5 = 29.

The same figure calculated for forests (with $S_{average}$ (0.012 ha) = 25):

- S_{forest}(0.01) = 20 + (24-25) = 20 – 1 = 19

And finally the figure for forest edges (with $S_{average}$ (0.005 ha) = 21):

- $S_{forest\ edge}$(0.01) = 42 + (24-21) = 42 + 3 = 45

2) *Calculation of* EDP *with the linear model*

The standardised species number per land use type is transformed into the characterisation factor *EDP* by applying equation [12]. Since EDP_{local_linear} is calculated using the relative species number a reference species number is necessary. This reference species number is the regional species number on 0.01 ha (e.g., 40 for the Swiss Plateau). The characterisation factor for agricultural land equals using average species number of the Swiss Plateau as reference:

- EDP_{local_linear} = 1 - (29/40) – 0.275.

For calculating the *EDP* with the nonlinear model equation [13] would work, where parameter *a* is 0.27 and parameter *b* is 1 (chapter 7.5.4).

3) Use of EDP *as a characterisation factor*

The calculated characterisation factor is then used to weight the surfacetime of land use. For example 12 ha of a field is used for 2 years to produce grain. The ecological damage is then

- D = 12 ha * 2 yr * 0.275 = 6.6 ha*yr.

This damage is interpreted as a weighted surfacetime according to its ecological damage potential. The basis for assessing the damage is the species diversity on the local scale.

Part D

Discussion and conclusion

11 Validity, reliability, and uncertainty

A model itself and the empirical data contained in it can be discussed with respect to uncertainty, validity, and reliability. The total uncertainty of a model has different components. The most important are incompleteness, model uncertainty, and data uncertainty (Funtowicz and Ravetz 1990, pp. 22). Additional types of uncertainty are distinguished and discussed by Morgan and Henrion (1990).

Nearly every model which deals with environmental problems is incomplete. This *incompleteness* results from either lack of knowledge about environmental mechanisms or the necessity to focus effort on a particular aspect of the problem. *Model uncertainty* refers to uncertainty about the functional relationships in the model and the numerical approximations of model parameters used. The choice of endpoints and indicators also contributes to this uncertainty. The *uncertainties of data* used in the model arise from the variability of environmental factors and inexactness of the measurements. A total model's data uncertainty can be assessed with a Monte-Carlo approach.

An objective in this project was to assess the impact of land use on ecosystem quality (EQ). Since EQ is a construct, changes in EQ cannot be measured directly. Indicators representing the construct EQ must be applied. Criteria of quality for such measurements are their validity and reliability.[13]

In general, *validity* refers to the ability to correctly measure what should be measured (Kromrey 1998). For example, when air tem-

[13] Measurement means the systematic (i.e., according to defined rules) assignment of numbers (or members of any cardinal or ordinal scale, encoded with symbols, colours or characters) to the characteristics of subjects in such a way that the relation between the numbers represents the relation between the subject's characteristics (adapted from Kromrey 1998, p. 225). Under this broad definition,

perature is measured with a thermometer, the result is valid. This is because, with our knowledge in physics, we can be sure that a thermometer measures temperature. However, it is more difficult to measure the quality of a product, because in this case quality depends on the consumer's needs. One might choose the wrong criteria, which do not indicate aspects of quality. Then the measurement is not valid, because quality was not what was measured.

Different aspects of validity can be distinguished.[14] *Semantic validity* (Kromrey 1998) is important if indicators are used for the measurement. This type of validity refers to the congruence between an indicator's intended meaning and the represented indicandum. In the case of correlated indicators (i.e., where in principle both indicators and indicandum can be measured and a correlation between them can be shown empirically) the semantic validity is high. Semantic validity is especially problematic for inference indicators (i.e., where the indicandum can only be measured via the indicator as a matter of principle). A direct measurement as previously shown in the case of ecosystem quality is not possible. The congruence of the intended meanings of the indicator and the indicandum can only be judged on the basis of plausibility and consistency.

Empirical validity includes both external and internal validity (Bortz 1993). Internal validity refers to the indisputable interpretation

---

one can measure the environmental awareness of people as well as the ecological burden of products.

14 Bossel (1994) splits up the validity of a model into different aspects. These are the validity of the model's behaviour and structure, i.e., it has to be proofed that the model represents the original system. In addition, he distinguishes empirical validity from the application validity. The former means that the results modelled correspond to empirical results and the latter means that goal and scope of the model meet the needs of the potential users. However, if inference indicators are used in a model, it is clearly not possible to assess empirical validity at all (in sensu Bossel), because there are no empirical results and the validity of model behaviour and structure can only be assessed in terms of plausibility and consistency.

of results. This generally tends to be high for laboratory experiments since one attempts to control all influencing variables. External validity refers to the ability to generalise results. This means that the results are not only valid given the conditions of the particular investigation, but that they also apply to other conditions (e.g., other regions, other climatic conditions, or other soil conditions).

Finally, the *reliability* of the measurement means that one gets a stable result for repeated measurements with constant conditions. Kromrey (1998) makes a distinction between the stability of the measurement across time, between observers, and between methods. For example, different observers might arrive at different results using the same device for measurements because of differences in their experience using this device. Then the measurement would not be reliable.

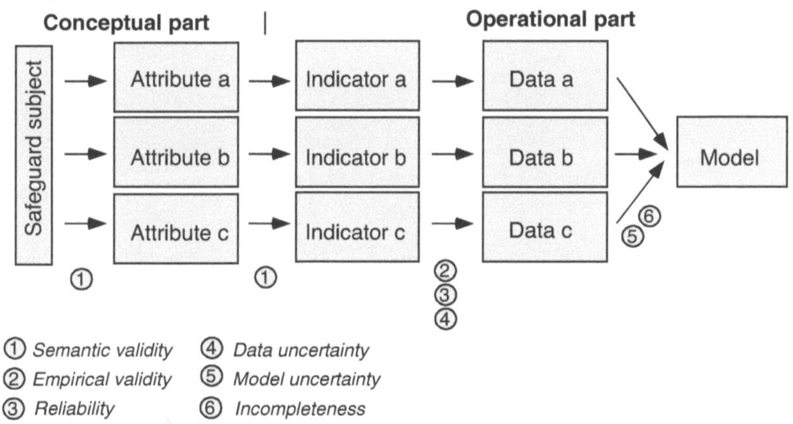

Fig. 11-1: Validity, reliability, and uncertainty in a model based on indicators.

The development of a model based on indicators is shown in Fig. 11-1 along with different types of validity, reliability and uncertainty. First, attributes representing the safeguard subject and indicators for measuring changes to the attributes are chosen. The choices affect semantic validity. Then, empirical data are sampled for the chosen indicators. In this stage empirical validity, reliability and data uncertainty apply. Finally, the empirical data is put into the model to compute the results. Here, model uncertainty and incompleteness are especially important, but different types of validity are as well.

12 Discussion of validity, reliability, and uncertainty of calculated EDPs

Model development has both a conceptual part and an operational part. I discuss the conceptual part in terms of chosen safeguard subject *ecosystem quality* and its attributes. The values the attribute *biodiversity* captures will be addressed in chapter 12.1. My discussion of the operational part includes the results of the local effect analysis (chapter 12.2), and the regional effect analysis (chapter 12.3), as well as modelling of the ecosystem damage potential (chapter 12.4).

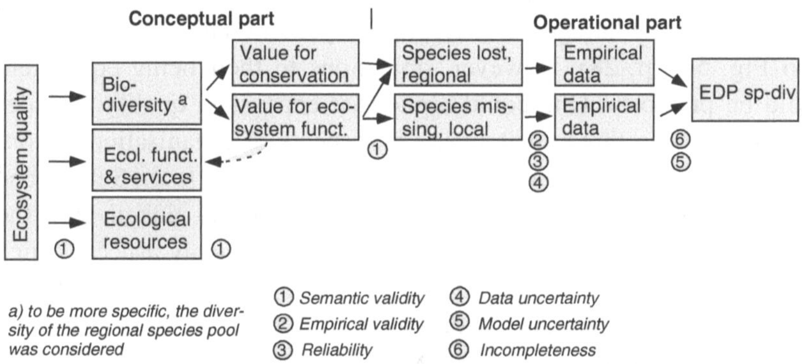

Fig. 12-1: Stages of model development. The goal is to develop the characterisation factor EDP$^{sp\text{-}div}$ (ecosystem damage potential for species diversity) expressing damage to the safeguard subject.

12.1 Chosen safeguard subject and its attributes

The safeguard subject which I have selected for consideration in the land use assessment is ecosystem quality EQ. For further modelling I have separated EQ into three attributes (biodiversity, ecosystem functions & services, ecological resources). These attributes cover different aspects of ecosystem quality from a systemic point of view (chapter 5.1.2, pp. 27).

An issue is whether or not the set of the three attributes is equivalent to the intended meaning of the safeguard subject. In terms of set theory different cases are possible. (i) The set of attributes is included in the intended meaning of the safeguard subject, but only captures a part of the meaning. (ii) The set of attributes matches exactly the intended meaning. (iii) The set of attributes covers more than the meaning of the safeguard subject.

In general, I think that equivalence between the set of attributes (biodiversity, ecosystem functions & services, and ecological resources) and the safeguard subject (ecosystem quality) is given (see also Fig. 5-2, p. 28). However, objections to their being equivalent could be put forth. One could argue that ecosystem quality includes more than these three attributes, e.g., aspects of organisation, autonomy, flexibility of development and buffer capacity (e.g., Scholz and Tietje in press). Some of these aspects are closely linked with the given attributes, but of course one could think of many aspects which should be considered in addition. The problem then is in trying to not choose too many attributes or only ones which are merely descriptive rather than normative. Another possible objection is to the attribute *ecological resources* being a part of ecosystem quality. One could argue that ecosystem quality should not include instrumental/resource values, because there is a safeguard subject resources. However, ecological resources like food and non-food biomass, surface water and fertile soil are components of the regional ecosystem and closely linked to the issue of ecological functionality and biodiversity. Therefore, they should be grouped here.

The conclusion is that the three attributes defined cover a large part of ecosystem quality, but of course, coming up with a definite answer about their semantic equivalence is not possible.

For the attribute *biodiversity* I distinguished its value for conservation and its value for ecosystem functioning (chapter 7.2, pp. 62). Biodiversity's value for ecosystem functioning obviously relates to the attribute *ecosystem functions & services*. Although this is better be assessed on the local scale rather than on the regional scale, a clear-cut assignment of type of value and spatial scale of assessment is difficult (see Fig. 12-1 and chapter 7.2). More value types could clearly be distinguished (chapter 5.1.3), but I have chosen only those which I made operable. Intrinsic values, for example, are not separated from use and non-use values (chapter 5.1.3, pp. 32). This is in line with international agreements that stress all of the value aspects of environmental goods, cf., Convention on Biological Diversity (UNEP 1992). In the field of Life Cycle Impact Assessment intrinsic and economic values of the natural environment are also acknowledged (Udo de Haes et al. 1999). However, some authors tend to limit the safeguard subject to its intrinsic value (chapter 6.6, pp. 51).

12.2 Land use and local species richness

Comparative assessment of land use types

In order to compare the different land use types, I calculated the standardised species number taking into account the area of the plot investigated. This is necessary, because as the area sampled increases, so does the species number (species-area relationship, see chapter 7.5.1, pp. 72). The results of the local effect analysis are given in chapter 7.5.4 (pp. 93). In general, the standardised species number per 0.01 ha ranges from 7 species (sport facilities) to 53 species (agricultural fallow with hedgerows). Based on local species richness, the following coarse classes can be distinguished (Tab. 7-5, p. 86):

- Species poor land use types (0-20 species/0.01 ha) are sport facilities, continuous urban area, conventional arable land, and intensive meadow. The reason for the low species number, and thus, its low ecological value, is the high-intensity of land use. A mixture of physical and chemical measures keeps the species number very low, which is partially intentional as in the case of agriculture. In contrast, there are land use types (heathland and peatbog) in this group which are very low in species number, but quite valuable from a nature conservation point of view. This is because in the last few decades these near-to-natural types became very rare in Europe, and thus, are now inhabited by threatened species (Bastian and Schreiber 1999, p. 311).[15] However, it can be expected that these land use types do not appear frequently in day-to-day LCA. Of course they do appear frequently in some cases, such as peatbog in an LCA of peat products (gardening and heating). Forest plantations were not explicitly considered in this study, although for practical LCA it is a quite important land use type. In general forest plantations are low in species richness because these plantations are very dense and dark. As a first approximation for its species richness 10 species/0.01 ha could be taken.

- Land use types with medium species richness (20-40 species/0.01 ha) are semi-natural broad-leafed forest, organic arable land, parts of industrial area with vegetation, green urban area, and natural grassland. This group is quite heterogeneous in terms of naturalness and land use intensity. For example, in green urban areas cultivation is generally quite intensive, whereas in semi-natural forests human impact is rarer and less intensively.
- Land use types with high species richness (40-60 species/0.01 ha) are industrial and agricultural fallow, organic meadow, forest edges, and agricultural fallow with hedgerows. This group

[15] Interestingly, heathland resulted from former land use activities. In former days heathland used to be public land and natural succession was prevented by the use of the upper layer of the heath as straw and by grazing of sheep (Ellenberg 1996, p. 719).

is also heterogeneous in terms of naturalness and land use intensity. One reason for the high species number is the low-intensity of land use. Agricultural fallow with hedgerows and forest edges are species rich, because these land cover types are heterogeneous in terms of abiotic conditions (e.g., light, wetness).

Semantic validity: Vascular plant species number as an indicator for species diversity?

The issue at this point is whether or not the chosen indicator is appropriate for indicating species diversity on the local scale. The indicator I used, species number per area, is quite simple. More sophisticated indicators exist (e.g., Shannon-Wiener index), which take species abundance into account as well (Magurran 1996). It can be assumed that the abundance of species has an impact on ecosystem functioning and, therefore, an indicator which measures species number and species abundance is preferable. However, due to lack of data more complex indicators could not be applied. Even in experimental settings the species number and the number of functional groups are used as indicators of local diversity (Hector et al. 1999).

Another issue is that vascular plants are a taxonomic group, which represents only a minor part of the total species richness. About 1.75 million species are described as existing on the global scale, but only 270'000 of these are plant species (Hawksworth and Kalin-Arroyo 1995, p. 118). However, three aspects indicate that vascular plant species make good indicators. First, the number of plant species described is quite close to the estimated number of total plant species (between 300'000 and 500'000). Second, Duelli and Obrist (1998) have shown empirically that the plant species number (Spermatophyta) correlates very well ($R^2 = 0.85$) with number of pooled species including plants, insects and spiders. However, the correlation may be weaker in other biogeographical regions. These results might not be applicable to the total species number if one takes also bacteria, molluscs, vertebrates, and other taxonomic groups into account. Third, as it is necessary to collect data on species numbers for very different land use types, the information which exists is sufficient for plant species only.

However on a qualitative level it is possible to roughly assess the semantic validity of the chosen indicator. The question is whether the results which I generated for vascular plant species can be generalised to other species groups:

- El Titi and Landes (El Titi and Landes 1990; El Titi 1991) compared conventional arable land use to less intensive arable land use on an experimental farm in Germany. Only ca. 25% of the nitrogen fertiliser and ca. 66% of pesticides which are usually applied in the conventional mode were used in the less intensive mode. They tracked the impact over a period of 10 years and found that soil fauna in less intensive arable land is richer and more active (e.g., 6 times more earthworms in terms of biomass).
- Similar results were found regarding insects and spiders by Pfiffner et al. (1995). Based on empirical findings from an experimental farm, about 2 times more species could be found on organically arable land compared to conventionally managed land. The difference was already evident in the first year of the experiment and continued for a period of three years.
- Frieben (1998) found that organic meadows are more species rich than conventionally managed meadows. However, she concluded that the plot's history in terms of land use is as important as the current mode of land use. She found two to three times more plant species on organically managed fields than on conventionally managed ones. In addition the organic fields were inhabited many threatened species.
- Flückiger (1999) conducted an investigation into forest edge biodiversity. He discovered that forest edges (42 species) are richer in vascular plant species than adjoining agricultural land

(24 species) or forests (20 species).[16] According to Flückiger, the same applies for insect and spider diversity. He found about 430 species at the forest edge, but only 120 in the middle of the forest.

- Based on a review of literature, Baur et al. (1997 p. 99) made a qualitative assessment of the relevance that agricultural land use types have for biodiversity. They found (i) that natural grassland, organic pastures, and meadows are *very valuable*; (ii) that less intensive meadows, organic orchards, and hedgerows are *valuable;* and that (iii) agricultural fallow and extensive fallow on former arable land are *less valuable*.

Based on this small literature review it can be concluded that my results generated for vascular plant species can be roughly generalised to other species groups. At least for agricultural land use types plant species richness and other species group's richness seems to be correlated.

Internal validity: indisputable interpretation of the findings

Species richness varies quite remarkably among the land use types. An important issue is whether these differences are a result of the different land use types or other uncontrolled variables.

The size of the area used is one variable that clearly has an influence on the species number. Therefore, I standardised the species number for the area using a double log species-area relationship. Other possible variables include the exposure of investigated plots (south orientation means more species), soil conditions, water conditions, and regional climate (Wohlgemuth 1998). All these variables influence species richness and contribute to the spread of the data. However, all data were taken from the Swiss Plateau and Germany which belong to the same biodiversity zone in sensu Barthlott (1998) and show comparable abiotic conditions on an European scale.

16 The standardised species numbers are calculated on pp. 153.

The number of plots investigated is another variable which can has influence on the species number assessed because it differs from land use type to land use type. Obviously, it is more appropriate to generalise the mean species number from 138 single observations (agricultural fallow) than the mean from 6 single observations (heathlands). In general, I focused on land use types which are important for agricultural products; and therefore, I sampled empirical findings about those in the first place.

Findings for semi-natural broad-leafed forests are problematic, since the plots were not chosen randomly; rather plots which are rich in species in a given forest were intentionally chosen. It can be expected that the species number in this forest type is actually a bit lower.

External validity: ability to generalise the results

The empirical data used in the effect analysis were acquired from Switzerland and Germany. It is disputable whether the results can be generalised to other regions or countries. Most of the data was sampled from regions which are used quite intensively with a large proportion of agriculture, forestry, and urban land. It can be expected that the absolute species number is valid for regions similar in land use intensity and biogeographical situation. However, I used a relative measure for local species richness and took the regional species richness as a reference. Therefore, the error might be less significant when the findings are generalised to other European countries. The coarse ranking of the land use types is expected to be stable across a wide geographical range.

Reliability: stability of measurements between different observers and between different methods

Reliability refers to the stability of empirical results when other researchers make the investigations or other methods are used. I considered many different data sources in the meta-analysis. In general, I assume that the reliability was quite high since most of the data was sampled according to the standardised method from Braun-Blanquet, which is applied in vegetation science. However, other methods were also used to determine species richness (chapter 7.5.1, pp. 72). This has an especially strong impact on the size and the number of the plots sampled. In general, with Braun-Blanquet lots of small plots are sampled, with the other method few large plots are investigated. The difference in the number of plots has also an impact on the external validity since the statistical basis differs. In addition, the accuracy of data can vary among observers, but it is difficult to judge this aspect of reliability on the basis of the available literature. Overall it can be concluded that reliability is rather good.

12.3 Land use in the region and species potentially lost

Regional effect analysis

The effect that different types of land use have on the regional scale potential loss of species since ~1850 was assessed with a regression analysis. A simple regression analysis results in the potential loss of a percentage of species as a consequence of a change in the percentage of specific land use types. The results of this analysis for 8 types of intensity of land use[17] were used in later stages of arriving at characterisation factors. Since a change in the proportion of one land use type implies that other land use types also change in proportion, I performed a multiple regression analysis to account for this corresponding change and to check the results of the individual regression.

[17] The land use types are:

- High-intensity (forest_hi) and low-intensity forest (forest_li)
- High-intensity (agri_hi) and low-intensity agricultural area (agri_li)
- Lakes and rivers (lakes) and terrestrial non-use area (non-use)
- High-intensity (artificial_hi) and low-intensity artificial surfaces (artificial_li)

The results of the individual regression are given in Fig. 7-18 (p. 117), Tab. 7-14 (p. 124). Based on the number of species potentially lost, following types of intensity of land use can be distinguished:

- Very detrimental land uses are both the high- and low-intensity artificial land use. Based on the linear model an increase in these types of intensity of 1% caused an increase in the potential species loss of approximately 1% to 4% in the region.
- Detrimental land uses are high-intensity agriculture and high-intensity forestry. Here the proportion of species potentially lost increase by about 0.4% to 0.8% when these types increase by 1%.
- Beneficial land uses are the low-intensity agriculture and low-intensity forestry. In this case the species loss decreases by about -0.2% to -1% when these types increase by 1%. This means that these types prevent species loss.

Semantic validity

The issue here is whether or not the chosen indicator (species number of vascular plants) is appropriate for indicating species diversity on the regional scale. Like on the local scale (chapter 12.2) species abundance could not be considered and more complex indicators were not possible. On this scale, however, I could perform the calculations taking all the vascular plant species observed in Switzerland into account, as well as focusing on those species which are threatened. The results are similar for both all the species (Tab. 7-15, p. 125) as well as those that are threatened (Tab. 7-16, p. 126) in terms of relative differences between land use types. However, for detrimental types of intensity (artificial_hi, artificial_li, forest_hi, agriculture_hi) the absolute percentage of species potentially lost is about twice as high for the threatened species than it is for all the species.

Internal validity

The issue of internal validity is whether or not the potential loss of species is a consequence of land use change. In the regression analysis

I have chosen to use the species potentially lost since ca. 1850 as dependent variable and the proportion of land use types observed in ca. 1970 as independent variable. The positive correlation I found between detrimental land use and species lost is, strictly speaking, only a coincidence. However, the issue is whether or not a functional relation is given. For this I have to assume (i) that land use intensity changed between ca. 1850 and present and (ii) that this change is, at least in part, responsible for the species potentially lost.

- In the case of agriculture, quite clearly a shift in mode from low-intensity land use to high-intensity one occurred during this time period. Average size of fields increased and agricultural landscape was optimised for intensive production. Since ca. 1950 chemical pesticides and artificial fertiliser particularly have been used in large amounts (Häberli et al. 1992). All these factors have resulted in a regional scale loss of biodiversity.
- Forest also became more intensive since ca. 1850 in Swiss Plateau. After this time, forests were clear-cut in large areas and reforested with plantations (mainly spruce) (Bürgi 1997, pp. 68). Formerly bright mixed forests became dark closed plantations consisting of trees of one age class. For example, Bürgi (1999) calculated an increase in the stock of growing forests in the canton Zurich from 180 m^3/ha in 1920 to about 350 m^3/ha in 1985. The dense plantations are quite detrimental to the silvicultural species diversity. These findings only apply partially to forests in alpine regions.
- Finally the intensity of artificial land use and its area (traffic infrastructure, housing, industrial plants etc.) has also increased since 1850 with negative consequences for biodiversity.

Korneck and Sukopp (1988, pp. 128) ranked 20 anthropogenic impacts on species diversity according to their severity for species diversity. They analysed the reason for each single species of the German red list for their threatening. The result was that land use is the most important determinant. Change in land use is according to them responsible for threatening 305 species. In contrast soil and air contamination threatens only 38 species.

At the same time the significant regressions in my investigation have an explainable variance of about 10% to 40%. This means that other independent factors (e.g., acidification, climatic changes) must have an impact on the quantity of species potentially lost in Switzerland as well. I did not take regressions with a slope of zero into account in my development of the characterisation factor, because this indicates that species lost [%] and the specific land use type [%] are independent. I tested this with F-statistics (cf., appendix 3).

External validity

An important issue of validity is whether the results generated for Switzerland can be generalised. Since the regression analysis was made for the 5 Swiss regions separately (Jura, Plateau, Northern Alps, Central Alps and Southern Alps), it is possible to assess the generality of the results. Focusing on the steepness (degree of slope parameter b) of the regression lines in the linear model for all the species, it is evident that the findings for some land use types (i.e., forest_hi, forest_li, agri_hi and agri_li) are quite similar from region to region (Fig. 7-18, p. 117). Regional differences were generated for artificial_hi and artificial_li. Generally speaking, the same applies for threatened species, but the values are more extreme.

However, a problem was that generally only a selection of the regions show significant regressions (Tab. 7-14 , p. 124 and Appendix 3). Taking artificial_hi for example significant results were generated for four regions. In contrast, for forest_hi the regressions were significant for only one region. The results for the plateau would be most appropriate for generalisation to other European countries, since this part of Switzerland is a typical non-alpine landscape which is used rather intensively (large proportion of agri_hi, forest_hi and artificial land use). However, quite surprisingly, no significant results were generated for agri_hi and forest_hi in this region. A possible explanation for this is that the sum of the proportions of various high-intensity land use is rather stable and that shifting between the two intensive land use types does not affect the number of species potentially lost.

For these reasons, I took the arithmetic mean of all significant regressions for each land use type into account in the development of the characterisation factor EDP. However, this restricts the ability to generalise the results, since the Jura and the alpine regions of Switzerland are not representative of other non-alpine European countries. The intensiveness of Swiss high-intensity agriculture for example would probably rate rather low in an international comparison.

Another qualitative regional scale assessment of very different land use types with respect to species and biotope conservation can be found in Kaule (1991; adapted by Bastian and Schreiber 1999, p. 318). He distinguished nine quality classes, which range from not valuable (class 1) to very valuable (class 9). Overall their assessment is in line with the ranking based on the analysis. However, major differences in the rankings of forests and agricultural areas exist. In Kaule's ranking forest_hi performs better than agri_hi and forest_li better than agri_li. The analysis I did resulted in a reverse ranking within each pair.

Reliability

The reliability of the results concerns its stability when changing observers or methods. This should be discussed especially in the case of empirical data used in regression analysis (i.e., species potentially lost and proportion of types of intensity, see pp. 83).

Regarding the species potentially lost (the dependent variable in the regression analysis), many people were involved in gathering the data. About 200 people with different degrees of expertise (professional botanists as well as lay people) generated the data of the 1967-1978 ground survey. For example is known that one layman produced extremely accurate data for the valley of the river Rhine (Wohlgemuth

1993). However, despite this variance, it can be assumed on a coarse scale that overall the data produced here is quite reliable.

For the historical data processed in the analysis the picture is quite different. Here the reliability is much lower, because the sources of historical data, varying in degree of accuracy, are mixed. For a part of the Swiss Plateau and the Jura literature sources are missing completely. I took this into account by focusing only on a part of the Swiss Plateau (pla-4). Regarding the Jura I did not exclude that part without literature data, since this part is rather small and the deviation in the percentage of species potentially lost, which were calculated only on basis of the data from herbaria, in comparison to the rest of the Jura seems to be small.

An issue which might affect the reliability is the assignment of the historical information mapped for large regions (Fig. 7-10, p. 98) to the small polygons of the Swiss EDV-Flora (Fig. 7-15, p. 112). This is a crucial step, which depends very much on the accuracy of spatial information given in the historical source. Some literature sources were excluded by Welten and Sutter (1982) according to Wohlgemuth (1993) because of missing spatial information, but large differences between the quality of the sources considered still exist. According to Wohlgemuth the historical number of species is dependent on the quality of the source, but overall he concludes that, at least in some regions, the historical information is useful for indicating the loss of species.

Another issue is a bias in the information contained in the herbaria, which together with the literature sources, completes historical information. More accurate species data are generally available for regions near the large cities, where universities supported the herbaria (Wohlgemuth 1993). This means that good data from herbaria are available for the regions with big cities. In this case it is difficult to decide whether the data indicates that large numbers of species are potentially lost in these regions because of (i) more accurate historical information or (ii) high-intensity land use detrimental for species diversity. Most likely it is a mixture of both, but in my opinion the second one has more impact.

With all the issues mentioned above in mind, the results of the spatial distribution of the species potentially lost (Fig. 7-15, p. 112) meet the expectations. Large numbers of species were lost in the agglomerations (Basel, Bern, Geneva, Lausanne, and Zurich), in regions with rather high-intensity agriculture (northern part of canton Bern, cantons Freiburg, Thurgau, and Waadt), and in the great valleys of the Alpine region. Of course this is only a rough test of plausibility, but even so, plausible results may indicate that the empirical data are reliable.

Another reliability issue is the calculation of the proportions of the eight types of intensity of land use for Switzerland. These calculations are based on the official Swiss area statistics and on the atlas of vegetation types produced by Hegg (1993). To calculate this, the joined land use type "arable land and meadows" from the Swiss area statistics was split apart (cf., appendix 2). The results of the calculation made in my project (i.e., 4154 km^2 of arable land and 4565 km^2 of meadows) are quite close to the areas published by the Swiss office of statistics (i.e., 4187 km^2 of arable land and 5270 km^2 of meadows) (BFS and BUWAL 1998, p. 34). To check the results for high and low-intensity forestry, I used the Swiss forestry inventory (WSL and BUWAL 1999). According to this inventory (page 105), there are 694 km^2 of forest with more than 90% presence of conifers and 672 km^2 with between 50% and 90% presence of conifers. Given the conditions of the Swiss plateau one can assume that the first type is completely high-intensity forest plantation and the second one partially so. According to the calculations I did, the area of forest plantations in the Swiss plateau is about 980 km^2. Based on these rough checks, one can conclude that the results for the land use proportions are rather reliable.

In order to check the results of the simple regression I also performed multiple regression (Tab. 7-13, p. 120) for each region as well as for all the species and for only those that are threatened. On a very general level the results are similar. Both methods showed that the proportion of artificial land to the species lost is positively correlated in every region. This land use type has a lot of explanatory power as an independent variable in the regression model.

High-intensity forestry and high-intensity agriculture have little explanatory power with either method (aside from high-intensity forests in the Jura). A possible explanation for this is that the polygons chosen (especially in the Alps) are very heterogeneous in terms of land use intensity. Maybe the ecological quality of a rather small proportion of a polygon is very good and could prevent species loss within the total area of the polygon of about 80 km^2. This would have a negative impact on the regression, since a polygon with a large proportion of high-intensity land use performs very well in terms of species loss (i.e., only a few species lost). At the same time, polygons were drawn by Welten and Sutter (1982) according to topographical conditions, which should prevent this effect to a certain degree. The range of altitude in meters within each polygon could be an indication of small and ecologically valuable areas in a region. Generally speaking, the greater the range in altitude, the more likely it is that there are valuable ecosystems remaining, which are difficult to use in an intensive mode. At the same time this means that polygons with a low range in altitude are more easy to use in an intensive mode. Wohlgemuth (1993) has shown empirically that this range, which varies from 100 meters to 1900 meters, and the species number observed are positively correlated for all polygons below the timberline ($species\_number = 0.174 * range\_in\_meters + 600$, $R^2=0.57$). It has not been tested yet whether the number of species lost and the altitude range are correlated as well. Nevertheless it is likely that polygons' large size negatively affects the regression analysis.

12.4 *Ecosystem damage potential ($EDP^{sp\text{-}div}$) for species diversity*

The objective of this project was to propose characterisation factors for different types of land use. These characterisation factors, which focus on species diversity, are called the *ecosystem damage potential* ($EDP^{sp\text{-}div}$). In order to calculate $EDP^{sp\text{-}div}$ I took into account the modelled effects on species diversity due to land use on both the local and regional scales. In addition, it was necessary to transform the observed effects into damage/benefits and to aggregate local and regional characterisation factors into a total factor.

An important issue for discussion is the validity of the characterisation factors $EDP^{sp\text{-}div}$. Model uncertainty and incompleteness of the model for calculating the characterisation factors are discussed below as well.

Validity of the characterisation factor EDP

Semantic validity poses the question of the intended meaning of the characterisation factor. My objective on the local scale was to indicate the value that species diversity has for ecosystem functioning. Therefore, I took not only the pure species number as an indicator, but performed a transformation which reflects the functional relationship between species number and ecosystem processes (chapter 7.5.3, pp. 91). On the regional scale, I distinguished between all of the species and those that are threatened in order to develop the characterisation factor. The choice of whether to consider all of the species or only the threatened ones has consequences with respect to the type of values which are taken into account. Focusing on the threatened species stresses the conservation value of species diversity. In contrast, taking all of the species includes also the value of species diversity for ecosystem functioning. This choice has also implications for the development of the total characterisation factor, which consists of a regional part and a local part. If one focuses on all of the species on the regional scale, the regional characterisation factor will be congruent with the local one, since the local characterisation factor also focuses on all of the species. This allows integration of local and regional scale results, based on an analytical procedure (chapter 7.7.1, p. 126, option 2). If one wishes to focus on the threatened species on the regional scale, quite different types of values would have to be addressed and this requires weighting of local and regional characterisation factors (chapter 7.7.1, p. 126, option 1).

One issue pertaining to *internal validity* is whether calculated factors only refer to land use impacts or also refer to other impacts. In the latter case, double counting of impacts would occur, since some other impact categories (e.g., nutrification and ecotoxicological effects) are assessed separately in LCIA. The characterisation factors for land use that I calculated integrate all the impacts which result from land use. These impacts include the intentional application of chemicals, such as in agriculture, where fertiliser and pesticides are used, as well as physical impacts like ploughing. However, it is necessary to distinguish (i) impacts on the plot of land use from (ii) impacts outside of this immediate plot resulting from runoff. In the former case, both the physical and chemical impacts of land use are included in the characterisation factor $EDP^{sp\text{-}div}$. However, in the latter case damage due to impacts from runoff are not included in the $EDP^{sp\text{-}div}$ factor, but must be considered in other impact categories. One can conclude from this that double counting is not a problem, when local $EDP^{sp\text{-}div}$ factors are used, which refer only to the plot in use. However, the regional factor includes also damages due to runoff, since species losses in a total region were assessed.

External validity motivates the question of whether the factors, calculated with Swiss and German data, can be generalised to a wider geographical range. In general the local characterisation factors are quite valid and can be used for other European countries with moderately intensive agriculture. This is because I used the relative species number in the characterisation factor, in which the average species number in the region serves as the reference. The coarse ranking of the land use types (cf., Tab. 7-7, p. 95) should be constant over a wide geographical range. It is plausible that in many countries, built up land and arable land perform very poorly, while organic meadow types and hedgerows perform very well. However, detailed ranking of the land use types will change when focusing on other European countries. In general the validity of the results will decrease with increasing distance to the study region. The regional characterisation factor, in contrast, is rather specific to the alpine region and should not be used for assessing land use damage outside of this region.

Model uncertainty

In general, *model uncertainty* refers to uncertainties in the functional relationships and in the parameters used in the model. This type of uncertainty is most significant when it occurs during the damage/benefits analysis. In this step, empirically determined effects are transformed into damages and benefits. The choice of a linear version of an effect-damage/benefits curve over a non-linear version implies uncertainty (e.g., Tab. 7-7, p. 95). Such uncertainty is reflected in the difference in values for *EDP* which occur from using linear transformation rather than non-linear transformation.

The use of relative species numbers also implies that there is a transition between damage and benefit (Fig. 7-21, p. 131). The local scale reference is the average species number in the region. Land use types with more species than the reference get negative $EDP^{sp\text{-}div}$ values, which means a benefit and vice versa. Of course it is debatable and uncertain, where to place the line between damage and benefit (see ranking in Tab. 7-7, p. 95, where for beneficial land use types this transition can be observed). However, characterisation factors are constructed in such a manner as to make the transition between values smooth. In addition, the ranking of the land use types is not influenced by changes to this border.

Another source of uncertainty pertains to the *aggregation of local and regional factors* into a total characterisation factor (Fig. 7-21, p. 131). Here different options for aggregation exist and the choice for one option implies uncertainty (chapter 7.7.1, pp. 126).

Incompleteness

An important issue for a characterisation factor is its incompleteness. In the case of land use, lack of knowledge about environmental mechanisms is rather moderate compared to other environmental impacts. For example, it is well known how agricultural land use affects biodiversity on the local scale. However, knowledge about the damage due to land use on the regional scale is rather small.

The most problematic component of incompleteness comes from the necessity to focus on a small part of the cause-effect chain (chapter 5.3.2, p. 37). This causes the model for calculating the characterisation factors to be rather incomplete.

In addition – and this issue is also related to semantic validity – only a part of the safeguard subject *ecosystem quality* was modelled, which is the species diversity. Species diversity, however, is one aspect of biodiversity, and biodiversity itself is one attribute of the safeguard subject *ecosystem quality*. In spite of the fact that a small part of the safeguard subject is covered, Life Cycle Impact Assessments are supposed to be more complete when considering the land use impacts on species diversity.

13 Discussion regarding the application of the method

13.1 Integrated damage/benefits of ecosystem quality versus changes in ecosystem quality

The characterisation factor or ecosystem damage potential ($EDP^{sp\text{-}div}$) refers to the potential damage (and potential benefits) of a specific land use type. Based on these factors, a specific land use activity (such as mining, with a transformation, an occupation, and a restoration phase) can be assessed in terms of two different aspects: first, the change of potential damages in time with respect to ecosystem quality and second, potential damages integrated over the total period of land use activity (chapter 7.4, pp. 66). These aspects should not be mixed up with land occupation or transformation. In both cases of land use, it is in principle possible to assess the change of potential damages in time and potential damages of land use integrated over time. This raises the question of which assessment aspect is more important. I argue that integrated damage should be assessed in the first place. This is because this gives basic information about the environmental performance of a land use activity. In addition, information about the changes of ecosystem quality that result from land use activity might be valuable. This information can include the delta change (initial compared to final), the direction of change (improvement or deterioration), and the extent of the amplitude. The extent of the amplitude could be seen as a kind of risk indicator (see Fig. 7-3b, p. 69). The higher the extent of the amplitude, the greater the risk that the ecosystem quality can reach a point of no return.

13.2 Use of generic characterisation factors

The generic characterisation factors proposed in this book can only be applied to marginal decisions; that is, decisions in which the consequences of a single decision are so small that the quality or quantity of environmental parameters of a region is not noticeably altered. However, many of those marginal decisions on a micro level can have substantial impact on the environment. I focused on this application type, because in general, LCA is a tool for supporting decisions on a micro level. In order to support decisions on a macro level (e.g., political decisions about pushing for less intensive agriculture) an average approach is advisable and the method developed here should not be used.

A set of characterisation factors in order to account for regional differences and for differences in species' degrees of endangerment is provided in this book. When applying the method described in this book I recommend differentiating between land use in alpine regions and in non-alpine regions. The regional $EDP^{sp\text{-}div}$ can only be used for assessing land use in alpine regions since the regression model, which is the basis for the regional $EDP^{sp\text{-}div}$, is very specific to Switzerland. Local *EDP's* are more generic and can be applied to the assessment of land use in Western European countries. However, it applies best to the regions where the empirical data comes from (i.e., Germany and the Swiss Plateau). Linear and non-linear models were used for the calculation of $EDP^{sp\text{-}div}$ factors. Though results are similar for both models, I recommend using the non-linear factors. This is because it is more appropriate to assume a logarithmic relation between the species lost and the damage than to assume a linear relation.

14 Conclusion

The purpose of this book is to improve contemporary Life Cycle Impact Assessment (LCIA) with the addition of assessment of land use. To accomplish this, an assessment method as well as generic information about the degree of ecosystem damage potentials of different land use activities were developed.

Application aspects

The characterisation factors $EDP^{sp\text{-}div}$ (ecosystem damage potential) are applied in assessing the damage of land use activities. Various factors are available, each for a specific land use type. Generic values for the factors are provided in this book (appendix 6a). Given the surfacetime of land use with the unit [ha*yr] and $EDP^{sp\text{-}div}$ factors, it is easy to calculate the potential damage, since it is the product of surfacetime and $EDP^{sp\text{-}div}$.

- *Inventory data*. For a potential user of the method outlined in this book, it is quite easy to calculate the potential damage of a land use. However, to do this, one must have data on the land use activities in the Life Cycle Inventory. This data includes the type of land use, the area used, and the period of usage. Depending on the goal and scope of the study, land use data can be sampled with varying degrees of accuracy. For example, data on background processes can be sampled on a coarse scale (on the level of the eight intensity types of land use) and data for central processes on a detailed scale (exact land use type).
- *Regionality*. The usefulness of differentiating inventory data about land use activities according to geographical regions depends very much on the spatial resolution of the LCIA. Though the present method is restricted to Switzerland, it shows how to take regional differences in the impact assess-

ment into account. It can be expected that future land use assessment methods will account for regional differences on an international scale. For this reason it is advisable, at least for long-term inventory activities, to register the geographical location of land use (e.g., country or region).

Modelling the ecosystem damage potential ($\text{EDP}^{\text{sp-div}}$*)*

In developing the characterisation factor $EDP^{sp\text{-}div}$, I focused on potential damage to biodiversity. In order to model this ecological damage, the local and regional scale were distinguished from one another.

- *Semantic validity*. For the operational calculations, only plant species diversity was taken into account, because of problems with data availability. However, plant species diversity is only one aspect of biodiversity, and this in turn, is only one attribute of ecosystem quality. While consideration of the diversity of species categories other than vascular plants (or other levels of biodiversity) would improve the assessment method, plant species diversity and other aspects of biodiversity minimally are not independent from one another. Keeping all the limitations in mind, present $EDP^{sp\text{-}div}$ factors at least provide additional insight into the ecosystem damage in a product's life cycle that results from land use. This alone has the potential to improve the quality of environmental decisions, since at present, not even land use (in terms of area and period of usage) is acknowledged in many decisions about products.
- *Uncertainties*. All the types of uncertainties (i.e., incompleteness, data uncertainty, and model uncertainty) in such a method of impact assessment are rather large. Data uncertainties were modelled with a Monte-Carlo approach. Other types of uncertainty are more difficult to address, although they are probably more important.

Further development

The method for Life Cycle Impact Assessment I developed allows to assess the ecological impact of land use in a product's life cycle with respect to species diversity. In addition other attributes of the ecosystem's quality (p. 27) should be taken into account in the impact assessment for land use. Further development of the method is therefore desirable. On the level of species diversity the focus of the impact assessment was very much on Switzerland and Germany. An important reason for this was the easy access to empirical information. It was quite difficult to access other countries' data sources. For a more complete assessment, however, species diversity data from many countries around the world would be necessary.

In addition it is necessary to develop the Life Cycle Inventory with respect to land use. This allows decision-makers to get access to land use data about processes in the product system (e.g., data about grey land use for extraction of raw materiels). But information about land use and about triggered ecosystem damages are not sufficient for ecological improvements. Hopefully, the awareness that land use is a hidden factor in many products and causes ecological damages will also raise, and that awareness will foster decisions to use land more ecologically.

Appendix 1
CORINE+ land-cover – Nomenclature definitions

Tab. App. 1-1: Classification of land-cover types according to (European Environmental Agency 2000). Additional types, which were not included in the original CORINE land-cover classification, are indicated by italic letters.

Type		Description (type of intensity)
1 Artificial surfaces		
10 Built up land		Land is totally covered with buildings or artificial infrastructure (artificial_hi)
11 Urban fabric	111 Continuous urban fabric	Most of the land is covered by buildings. Roads and artificially surfaced area cover almost all the ground. Non-linear areas of vegetation and bare soil are exceptional. At least 80% of the total area are sealed. (artificial_hi)
	112 Discontinuous urban fabric	Most of the land is covered by structures. Buildings, roads and artificially surfaced areas associated with areas with vegetation and bare soil, which occupy discontinuous but significant surfaces. Less than 80% of the total area are sealed. (artificial_hi)
	113 Urban fallow	Remains of demolished houses are removed and areas are levelled. (non-use)
	114 Rural settlement	Village with buildings, streets and gardens (artificial_li)
12 Industrial, commercial and transport	121 Industrial or commercial units	Artificially surfaced areas (with concrete, asphalt, tamacadam, or stabilised, e.g., beaten earth) devoid of vegetation occupy most of the area in question, which also contains buildings and/or areas with vegetation. (artificial_hi)
	1211 Industrial area built up part	Land is totally covered with industrial buildings or infrastructure (artificial_hi)
	1212 Industrial area part with vegetation	High or medium proportion of sealed area; open soils, sometimes contaminated with chemicals. (artificial_hi)
	122 Road and rail networks and associ-	Motorways, railways, including associated installations (stations, platforms, embank-

	ated land	ments). Minimum width to include: l 00 m.
	1221 *Road networks*	(artificial_hi)
	1222 *Road embankments and associated land*	(artificial_li)
	1223 *Rail networks*	(artificial_hi)
	1224 *Rail embankments and associated land*	(artificial_li)
	1225 *Rail fallow*	Fallow land with spontaneous vegetation. (non-use)
	123 Port areas	Infrastructure of port areas, including quays, dockyards and marinas. (artificial_li)
	124 Airports	Airport installations: runways, buildings and associated land. (artificial_li)
	125 Industrial fallow	Areas with remains of industrial buildings; deposits of rubble, gravel, sand and industrial waste. (non-use)
13 Mine, dump and construction sites	131 Mineral extraction sites	Areas with open-pit extraction of industrial minerals (sandpits, quarries) or other minerals (opencast mines). Includes flooded gravel pits, except for river-bed extraction. (artificial_hi)
	132 Dump sites	Landfill or mine dump sites, industrial or public. (artificial_hi)
	133 Construction sites	Spaces under construction development, soil or bedrock excavations, earthworks. (artificial_hi)
	134 Mining fallow	Levelled deposits of slag heaps, rubble, gravel and sand. (non-use)
14 Artificial, non-agricultural areas with vegetation	141 Green urban areas	Areas with vegetation within urban fabric. Includes parks and cemeteries with vegetation. (artificial_li)
	142 Sport and leisure facilities	Camping grounds, sports grounds, leisure parks, golf courses, racecourses, etc. Includes formal parks not surrounded by urban zones. (artificial_li)
2 Agricultural areas		
21 Arable land		Cultivated areas regularly ploughed and generally under a rotation system.

	211 Non-irrigated arable land	Cereals, legumes, fodder crops, root crops and fallow land. Includes flower and tree (nurseries) cultivation and vegetables, whether open field, under plastic or glass (includes market gardening). Includes aromatic, medicinal and culinary plants. Excludes permanent pastures.
	2111 Intensive	Chemical-synthetic and organic fertiliser as well as pesticides are applied. (agri_hi)
	2112 Integrated	Chemical-synthetic and organic fertiliser as well as pesticides are applied. However, the input of these substances is reduced. (agri_hi)
	2113 Organic	Chemical-synthetic fertiliser and pesticides are not allowed, organic fertiliser is applied. (agri_hi)
	2114 Fibre/energy crops	*Crops for fibre and energy production: 21141 Kenaf (Hibiscus cannabinus L.), 21142 Hemp (Cannabis sativa L.), 21143 Chinese reed (Miscanthus sp.), also 21144* energy forests with short rotation time (agri_hi)
	2115 Agricultural fallow	(non-use)
	212 Permanently irrigated land	Crops irrigated permanently and periodically, using a permanent infrastructure (irrigation channels, drainage network). Most of these crops could not be cultivated without an artificial water supply. Does not include sporadically irrigated land. (agri_hi)
	213 Rice fields	Land developed for rice cultivation. Flat surfaces with irrigation channels. Surfaces regularly flooded. (agri_hi)
22 Permanent crops		Crops not under a rotation system which provide repeated harvests and occupy the land for a long period before it is ploughed and replanted: mainly plantations of woody crops. Excludes pastures, grazing lands and forests.
	221 Vineyards	Areas planted with vines.
	2211 Intensive	(agri_hi)
	2212 Organic	(agri_li)
	222 Fruit trees and berry plantations	Parcels planted with fruit trees or shrubs: single or mixed fruit species, fruit trees associated with permanently grassed sur-

		faces. Includes chestnut and walnut groves.
	2221 Intensive orchards	Orchards with small growing fruit trees (agri_hi)
	2222 Organic orchards	Orchards with meadows and large fruit trees (agri_li)
	223 Olive groves	Areas planted with olive trees, including mixed occurrence of olive trees and vines on the same parcel.
23 Pastures *and meadows*	231 Pastures *and meadows*	Dense, predominantly graminoid grass cover, of floral composition, not under a rotation system. Mainly used for grazing, but the fodder may be harvested mechanically. Includes areas with hedges (bocage)., e.g., oat grass meadow (Arrhenatherion, Polygono), fertilised moist meadow (Calthion)
	2311 Intensive pasture and meadows	Meadows mechanically harvested 3 times or more per year, fertiliser applied, perhaps on former arable land (agri_hi)
	2312 Less intensive pasture and meadows	Meadows mechanically harvested 2 or 3 times per year, reduced input of fertiliser, perhaps on former arable land (agri_li)
	2313 Organic pasture and meadows	Meadows mechanically harvested 1 time per year, no fertiliser, perhaps on former arable land (agri_li)
24 Heterogeneous agricultural areas	241 Annual crops associated with permanent crops	Non-permanent crops (arable lands or pasture) associated with permanent crops on the same parcel.
	242 Complex cultivation	Juxtaposition of small parcels of diverse annual crops, pasture and/or permanent crops.
	243 Land principally occupied by agriculture	Areas principally occupied by agriculture, interspersed with significant natural areas. (agri_li)
	244 Agro-forestry areas	Annual crops or grazing land under the wooded cover of forestry species. (agri_li)
	245 Agricultural fallow with hedgerows	(non-use)
3 Forests and semi-natural areas		
31 Forests	311 Broad-leafed forest	Vegetation formation composed principally of trees, including shrub and bush understories, where broad-leafed species predominate. (Presence of conifers 0-10%)

	3111 Broad leafed plantations	Plantations of fast growing tree species like popular (forest_hi)
	3112 Semi-natural broad-leafed forests	Natural or semi-natural forests, where coniferous species predominate, either moist or arid form (forest_li)
	312 Coniferous forest	Vegetation formation composed principally of trees, including shrub and bush understories, where coniferous species predominate. (Presence of conifers 91-100%)
	3121 Coniferous plantations	Plantations of fast growing tree species like Picea abies (forest_hi)
	3122 Semi-natural coniferous forests	Natural or semi-natural forests, where coniferous species predominate (forest_li)
	313 Mixed forest	Vegetation formation composed principally of trees, including shrub and bush understories, where broad-leafed and coniferous species co-dominate. (forest_li)
	3131 Mixed broad-leafed forest	Presence of conifers 11-50% (forest_li)
	3132 Mixed coniferous forest	Presence of conifers 51-90% (forest_li)
	314 Forest Edge	Borderline between silvicultural and agricultural area (non-use)
32 Shrub and/or herbaceous vegetation associations	321 *Semi*-Natural grassland	Low productivity grassland. Often situated in areas of rough uneven ground. Frequently includes rocky areas, briars, and heathland., e.g., xeric grassland (Xero-Bromium), mesoxeric grassland (Meso-Bromium), mat-grass pasture (Violion), moor-grass meadow (Molinion) (agri_li)
	322 ~~Moors and~~ heathland	Vegetation with low and closed cover, dominated by bushes, shrubs and herbaceous plants (heath, briars, broom, gorse, laburnum, etc.). (non-use)
	323 Sclerophyllous vegetation	Bushy sclerophyllous vegetation. Includes maquis and garrige. Maquis: a dense vegetation association composed of numerous shrubs associated with siliceous soils in the Mediterranean environment. Garrigue: discontinuous bushy associations of Mediterranean calcareous plateaus. Generally composed of kermes oak, arbutus, lavender, thyme, cistus, etc. May include a few isolated trees. (non-use)
	324 Transitional	Bushy or herbaceous vegetation with scat-

	woodland/shrub	tered trees. Can represent either woodland degradation or forest regeneration/ colonisation. (non-use)
	325 Hedgerows	Bushy or herbaceous vegetation with scattered trees at the edge of fields (non-use)
33 Open spaces with little or no vegetation	331 Beaches, dunes, and sand plains	Beaches, dunes and expanses of sand or pebbles in coastal or continental areas, including beds of stream channels with torrential regime. (non-use)
	332 Bare rock	Scree, cliffs, rocks and outcrops. (non-use)
	333 Sparsely vegetated areas	Includes steppes, tundra and badlands. Scattered high-attitude vegetation. (non-use)
	334 Burnt areas	Areas affected by recent fires, still mainly black. (non-use)
	335 Glaciers and perpetual snow	Land covered by glaciers or permanent snowfields. (non-use)
4 Wetlands		
41 Inland wetlands		Non-forested areas either partially, seasonally or permanently waterlogged. The water may be stagnant or circulating. (non-use)
	411 Inland marshes	Low-lying land usually flooded in winter, and more or less saturated by water all year round. (non-use)
	412 Peatbogs	Peatland consisting mainly of decomposed moss and vegetable matter. May or may not be exploited. e.g., raised bog (Sphagnetum), intermediate mires (e.g., Scheuchzerietum) and fens (Caricetum) (non-use)
42 Coastal wetlands		Non-wooded areas either tidally, seasonally or permanently waterlogged with brackish or saline water.
	421 Salt marshes	Low-lying areas with vegetation, above the high-tide line, susceptible to flooding by sea water. Often in the process of filling in, gradually being colonised by halophilic plants.
	422 Salines	Salt-pans, active or in process. Sections of salt marsh exploited for the production of salt by evaporation. They are clearly distinguishable from the rest of the marsh by their segmentation and embankment systems.
	423 Intertidal flats	Generally unvegetated expanses of mud,

		sand or rock lying between high and low water-marks. On contour on maps.
5 Water bodies		
51 Inland waters	511 Water courses	Natural or artificial water-courses serving as water drainage channels. Includes canals. Minimum width to include: 100 m.
	512 Water bodies	Natural or artificial stretches of water.
	5121 Artificial lakes	Reservoir in a valley because of damming up rivers.
	5122 Natural lakes	Natural water bodies frequently because of former existence of glaciers.
52 Marine waters	521 Coastal lagoons	Unvegetated stretches of salt or brackish waters separated from the sea by a tongue of land or other similar topography. These water bodies can be connected with the sea at limited points, either permanently or for parts of the year only.
	522 Estuaries	The mouth of a river within which the tide ebbs and flows.
	523 Sea and ocean	Zone seaward of the lowest tide limit.

Appendix 2
Proportion of land use types in Swiss regions

Tab. App. 2-1: Proportions [%] of land use types in the Swiss regions Jura (jur), Plateau (pla), Northern Alps (n-alp), Central Alps (c-alp) and Southern Alps (s-alp). Grey shadowed areas show the proportions for the detailed partitioning of the Swiss plateau. Total area of Switzerland is 41128 km^2.

land use type	jur	pla	pla-2a	pla-2b	pla-3	pla-4	n-alp	c-alp	s-alp	above tim-berl.	water	Swiss Ø
Forest												
Closed forest *(high- intensity closed forest)*	7.0	9.2	13.5	3.0	9.9	8.2	3.9	3.1	22.8	0.2	0.2	5.1
Closed forest *(low-intensity closed forest)*	36.8	15.4	21.3	12.4	13.5	15.9	30.6	27.3	34.6	3.4	1.8	19.6
Open forest	1.2	0.0	0.0	0.1	0.0	0.0	2.8	7.1	3.7	0.9	0.3	2.0
Shrub forest	0.0	0.0	0.0	0.0	0.0	0.0	1.7	6.0	4.2	1.2	0.0	1.6
Hedgerows, cluster of trees	3.5	2.1	2.3	2.8	2.4	1.6	4.6	5.1	3.7	1.1	0.5	2.9
Agriculture area												
Vineyards	0.3	0.6	0.7	5.6	0.1	0.0	0.1	0.7	1.0	0.0	0.0	0.3
Intensive fruit trees	0.2	0.7	0.4	1.4	1.2	0.2	0.1	0.5	0.0	0.0	0.0	0.3
Extensive fruit trees	1.4	2.7	2.0	1.7	3.8	2.0	0.6	0.2	0.1	0.0	0.0	1.0
Horticulture	0.0	0.2	0.3	0.5	0.2	0.2	0.0	0.0	0.1	0.0	0.0	0.1
Arable land and meadow *(arable land)*	10.5	29.3	23.3	33.8	25.3	34.2	5.4	3.4	2.1	0.1	0.3	10.1
Arable land and meadow *(intensive meadow)*	9.4	18.2	15.0	9.4	20.8	18.3	9.0	2.5	3.8	0.2	0.4	8.0
Arable land and meadow *(organic meadow)*	8.6	4.3	1.7	7.4	3.8	5.1	3.9	2.0	0.7	0.2	0.1	3.1

land use type	jur	pla	pla-2a	pla-2b	pla-3	pla-4	n-alp	c-alp	s-alp	above timberl.	water	Swiss Ø
Extensive vineyards	0.0	0.0	0.0	0.0	0.0	0.0	0.0	0.0	0.2	0.0	0.0	0.0
Heimweiden	1.1	1.6	0.5	1.1	1.9	1.9	2.4	0.7	0.5	0.0	0.0	1.2
Maiensässen, Heualpen, Bergwiesen	0.0	0.0	0.0	0.0	0.0	0.0	1.8	2.5	0.8	0.3	0.0	0.8
Pastures in the Alps and the Jura	12.9	0.1	0.0	0.2	0.1	0.2	18.4	15.0	3.6	29.5	0.8	14.5
Non-productive area												
Lakes	0.3	0.4	0.1	0.4	0.5	0.5	1.4	0.8	0.8	0.3	94.0	3.5
Running waters with embankment	0.2	0.8	1.7	0.6	0.7	0.7	1.0	1.5	1.4	0.2	0.1	0.7
Non-productive vegetation	0.2	0.3	0.2	0.2	0.5	0.3	2.2	5.6	3.1	2.1	0.4	1.9
Area without vegetation	0.2	0.0	0.0	0.0	0.0	0.0	5.7	12.7	3.7	59.9	0.1	17.5
Artificial surfaces												
Buildings – built up part	0.9	1.9	2.4	2.3	2.2	1.5	0.7	0.4	1.6	0.1	0.1	0.8
Buildings – part with vegetation	2.4	5.1	5.8	8.5	5.6	3.7	1.4	0.9	3.6	0.0	0.3	2.1
Industrial plants – built up part	0.1	0.3	0.5	0.3	0.4	0.2	0.1	0.0	0.2	0.0	0.0	0.1
Industrial plants – part with vegetation	0.2	0.6	0.9	0.7	0.6	0.5	0.2	0.1	0.4	0.0	0.0	0.2
Special artificial surfaces	0.4	0.9	1.4	1.1	0.8	0.7	0.3	0.4	0.5	0.0	0.1	0.4
Sport-, leisure and green urban areas	0.3	0.8	1.1	1.2	0.9	0.6	0.2	0.1	0.5	0.0	0.2	0.3
Road networks	1.9	3.8	4.4	4.9	4.3	3.0	1.5	1.2	2.1	0.1	0.2	1.7
Rail networks	0.2	0.4	0.6	0.5	0.4	0.3	0.2	0.1	0.3	0.0	0.1	0.2
Airports	0.0	0.1	0.0	0.2	0.2	0.0	0.0	0.0	0.0	0.0	0.0	0.0
Total	100.0	100.0	100.0	100.0	100.0	100.0	100.0	100.0	100.0	100.0	100.0	100.0

Tab. App. 2-2

Type of intensity	jur	pla	pla-2a	pla-2b	pla-3	pla-4	n-alp	c-alp	s-alp	above timberl.	water	Swiss Ø
forest_hi	7.0	9.2	13.5	3.0	9.9	8.2	3.9	3.1	22.8	0.2	0.2	5.1
forest_li	41.5	17.5	23.5	15.3	15.9	17.6	39.6	45.5	46.1	6.7	2.6	26.1
agri_hi	20.4	48.9	39.7	50.7	47.6	52.8	14.6	7.2	7.0	0.3	0.7	18.7
agri_li	23.9	8.8	4.1	10.4	9.6	9.3	27.2	20.3	5.9	30.1	1.1	20.5
lakes	0.5	1.3	1.9	1.0	1.1	1.3	2.4	2.2	2.2	0.5	94.1	4.2
non-use	0.4	0.4	0.2	0.2	0.5	0.3	7.9	18.3	6.8	62.0	0.5	19.4
artificial_hi	3.5	7.5	9.3	9.2	8.3	5.8	2.7	2.2	4.8	0.2	0.4	3.3
artificial_li	2.9	6.5	7.8	10.3	7.1	4.8	1.8	1.1	4.5	0.0	0.5	2.6
total	100.0	100.0	100.0	100.0	100.0	100.0	100.0	100.0	100.0	100.0	100.0	100.0

Proportion of types of intensity of land use in Switzerland

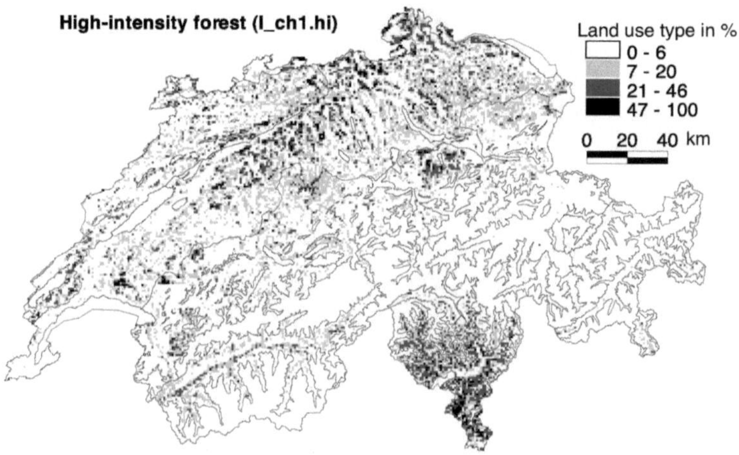

Fig. App. 2-1

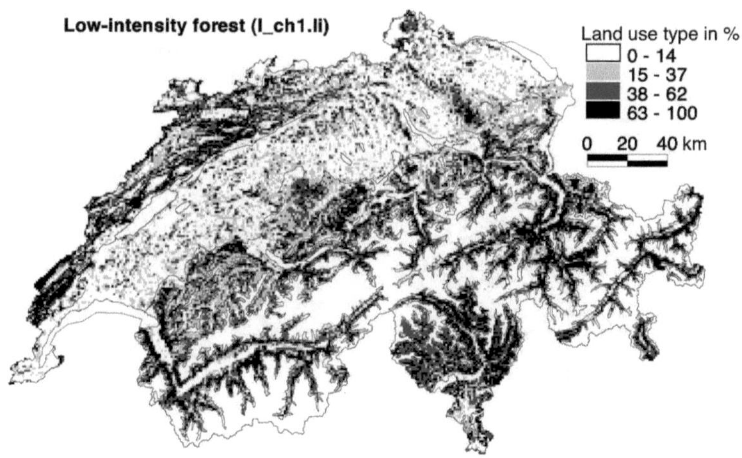

Fig. App. 2-2

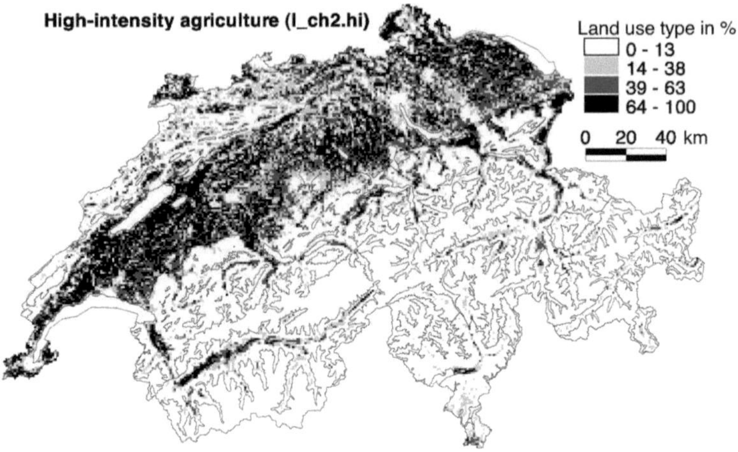

Fig. App. 2-3

Fig. App. 2-4

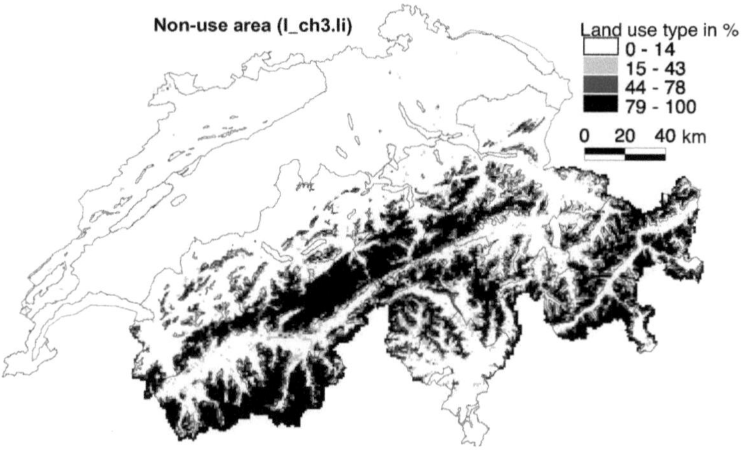

Fig. App. 2-5

Fig. App. 2-6

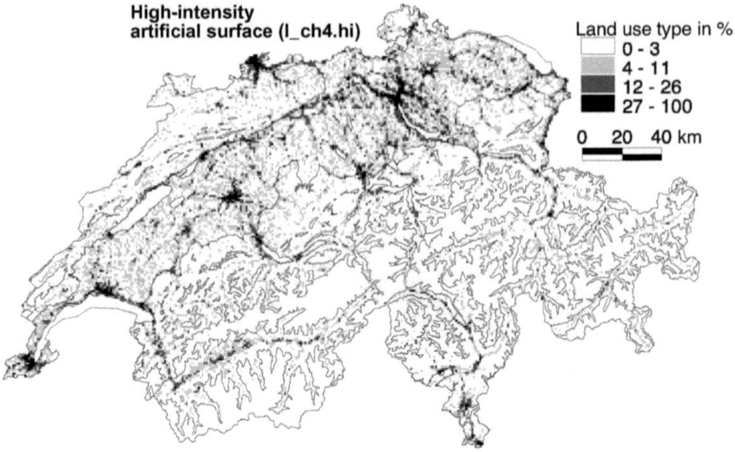

Fig. App. 2-7

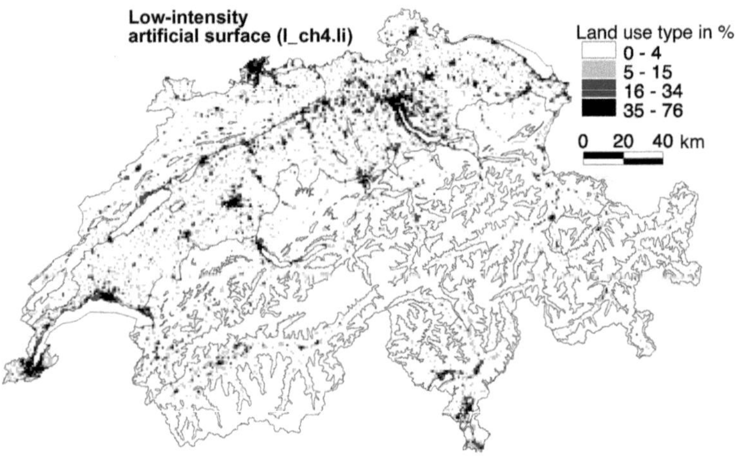

Fig. App. 2-8

Appendix 3
Detailed results of the regression analysis on the regional scale

Linear regression of proportion of land use types (independent variable) and the species, which are potentially lost as dependent variable. Each group of regressions is for one region (e.g., the first for the Swiss region "Jura").

The land use types considered: high-intensity forest (forest_hi), low-intensity forest (forest_li), high-intensity agricultural area (agri_hi), low-intensity agricultural area (agri_li), lakes and rivers (lakes), terrestrial non-use area (non-use), high-intensity artificial surfaces (artificial_hi) and low-intensity artificial surfaces (artificial_li). The regressions for the threatened species are not shown.

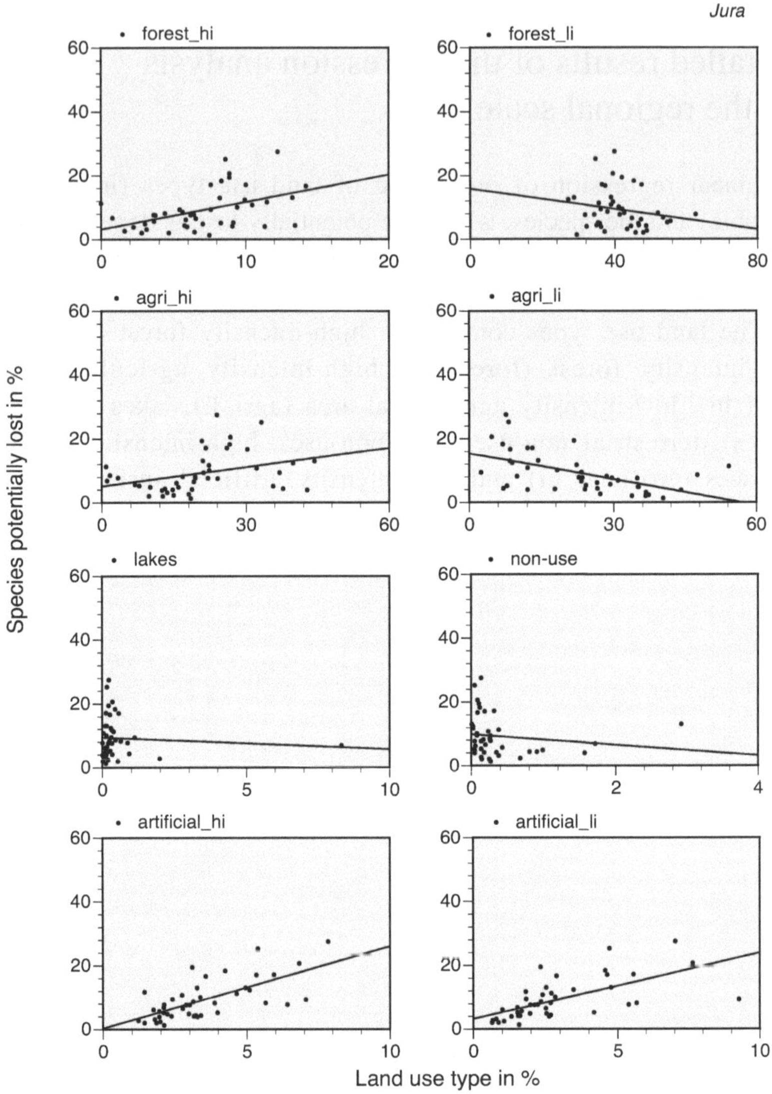

Fig. App. 3-1: Linear regression for Jura.

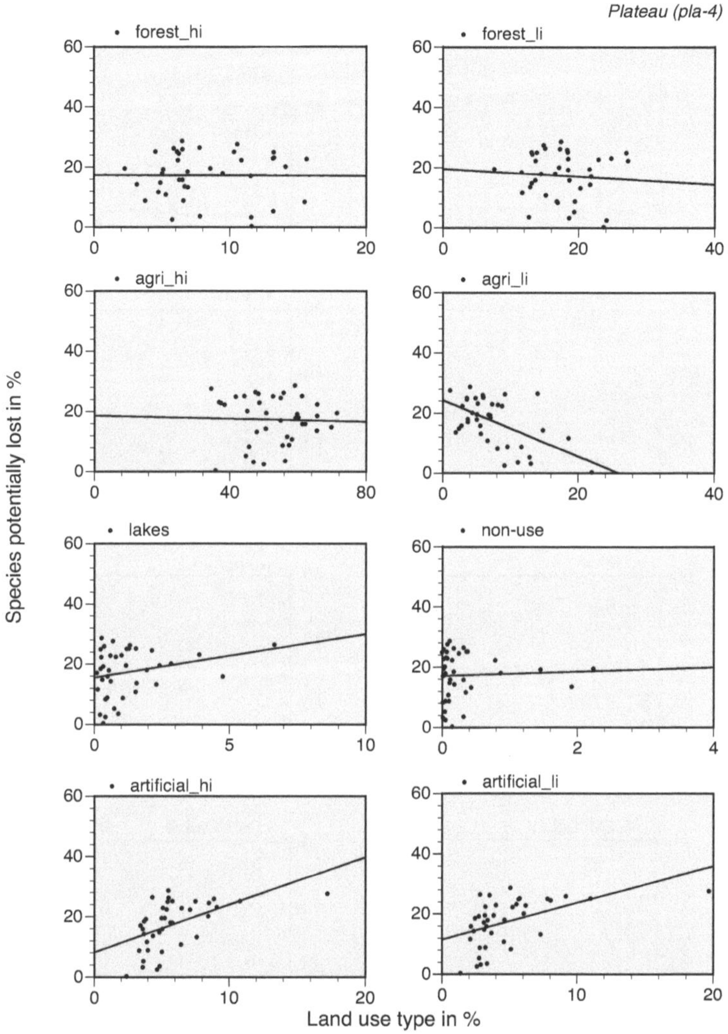

Fig. App. 3-2: Linear regression for Plateau (pla-4).

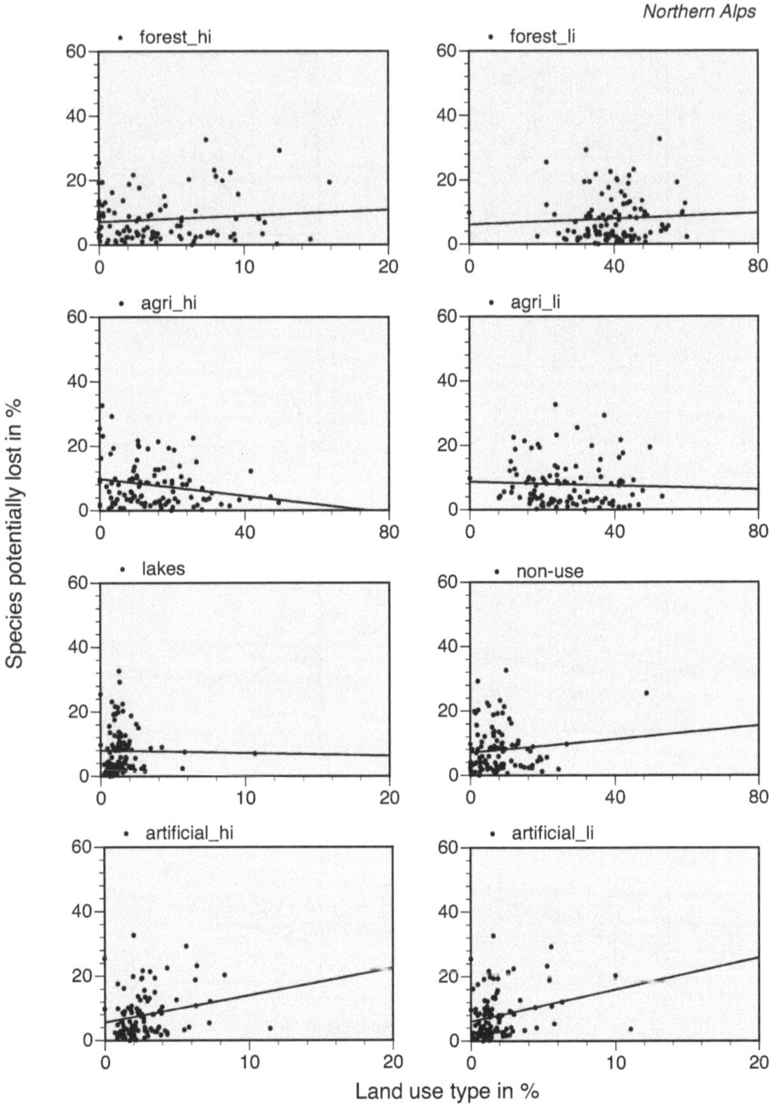

Fig. App. 3-3: Linear regression for Northern Alps.

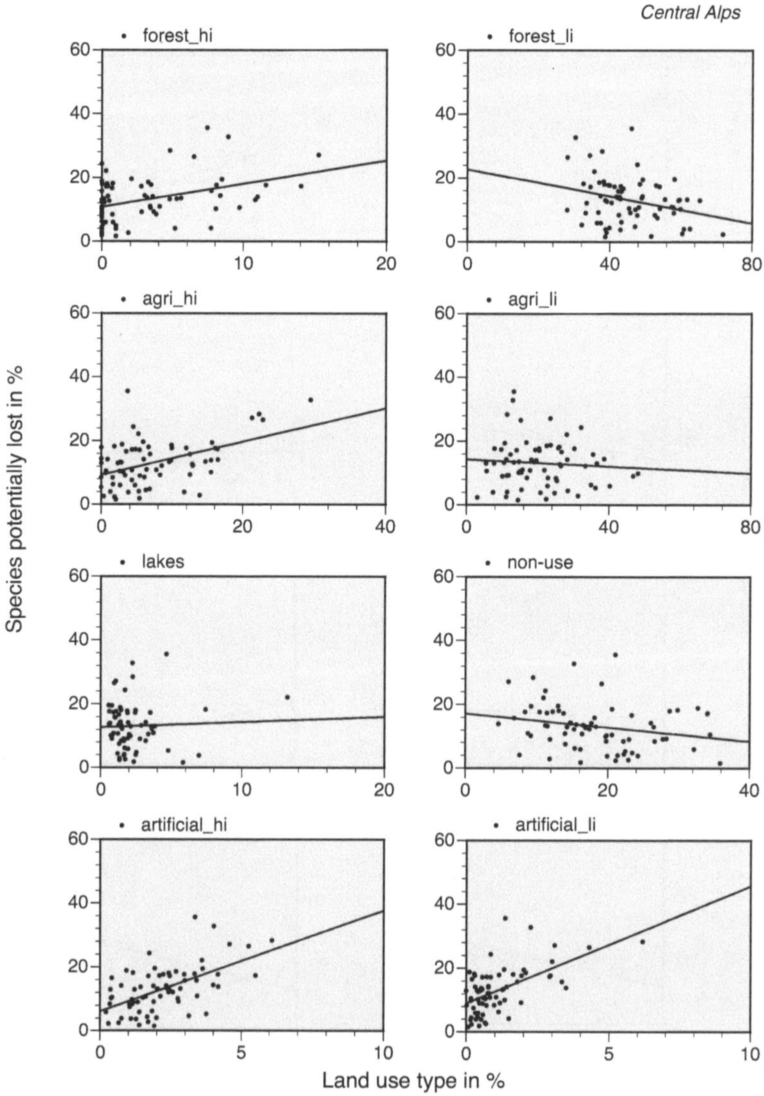

Fig. App. 3-4: Linear regression for Central Alps.

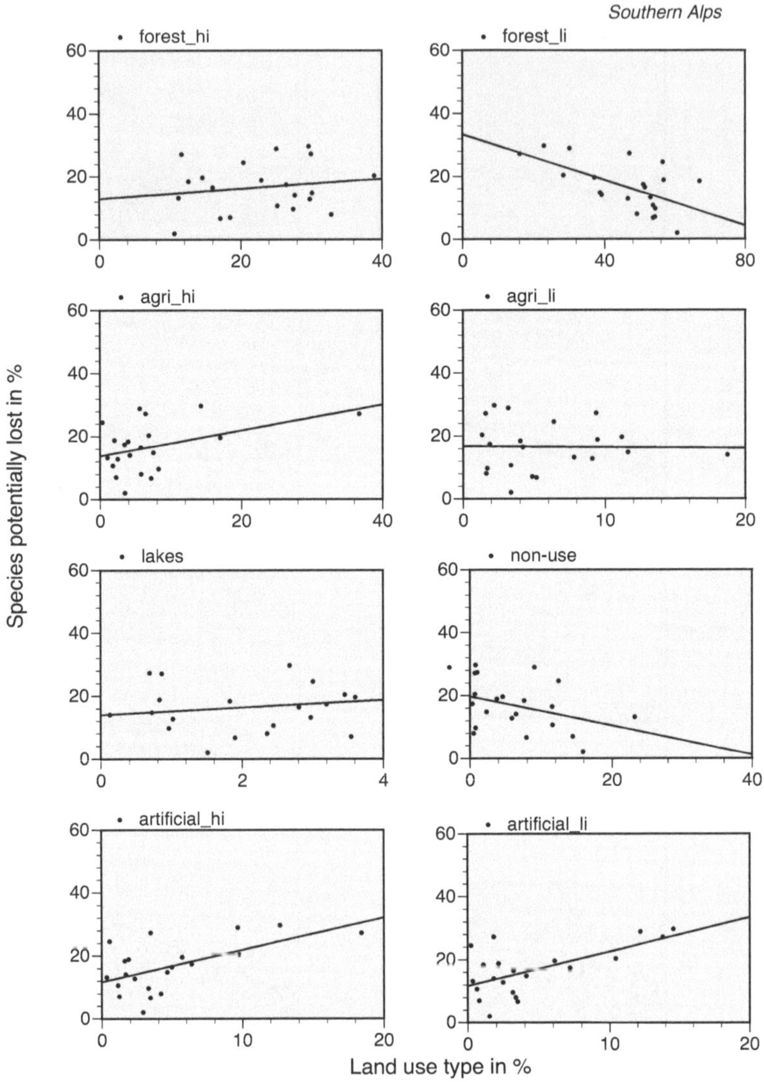

Fig. App. 3-5: Linear regression for Southern Alps.

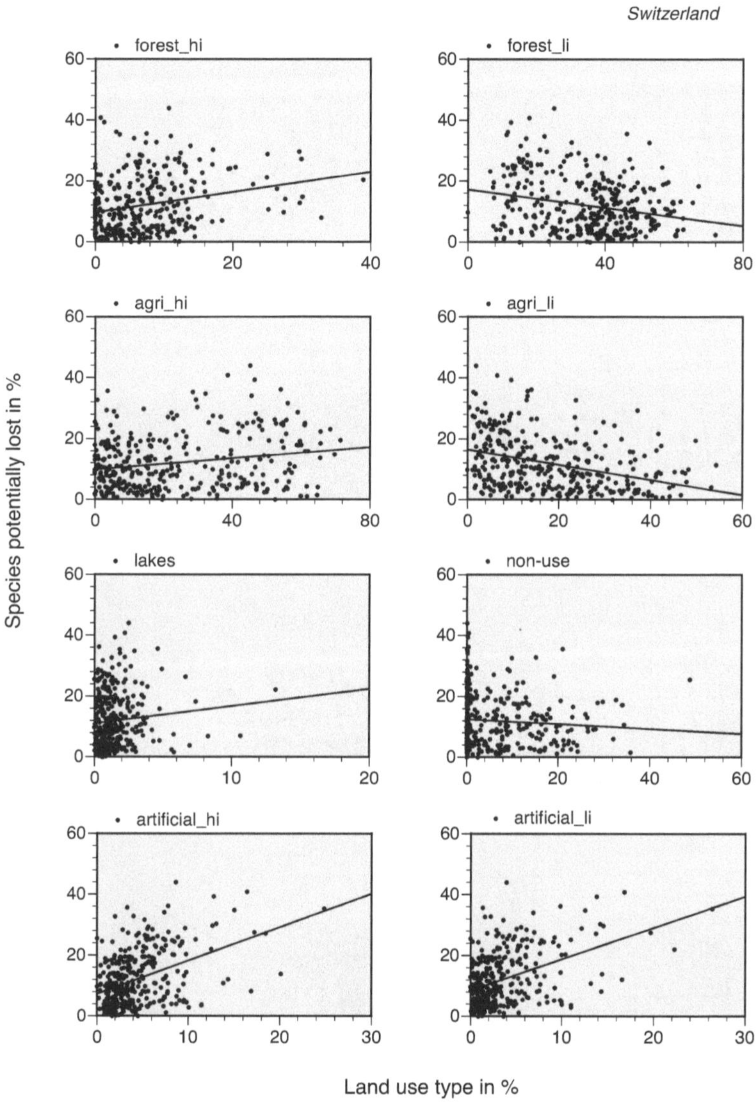

Fig. App. 3-6: Linear regression for total Switzerland.

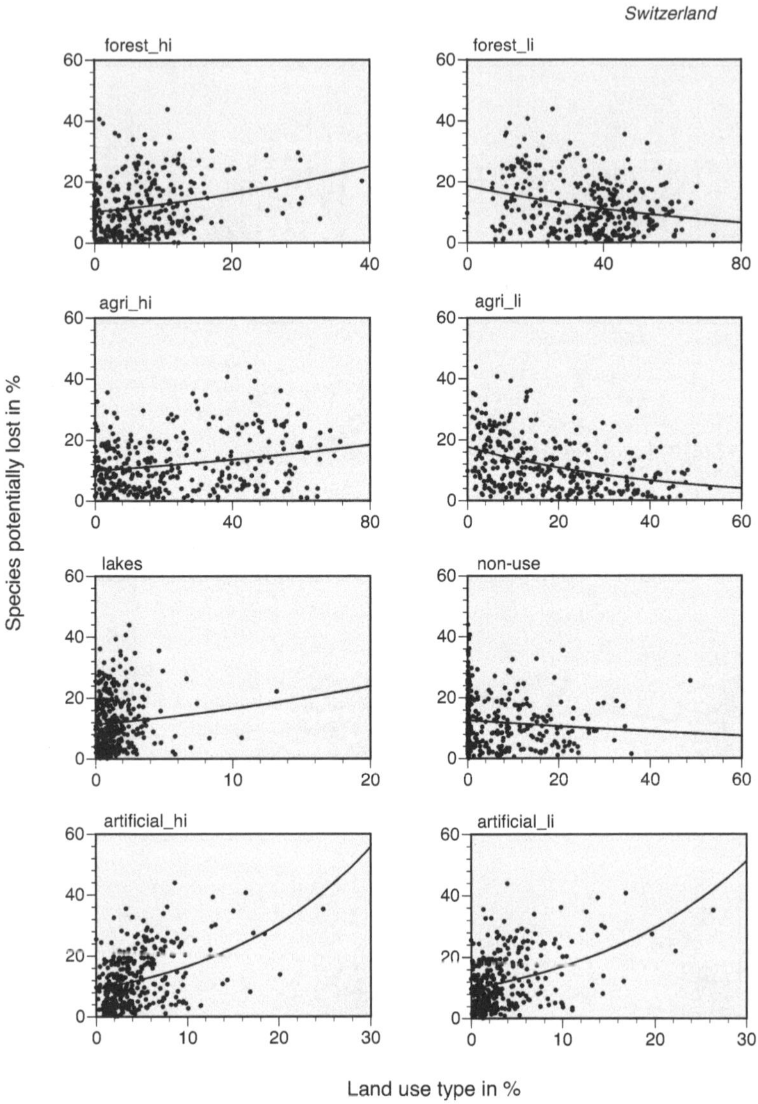

Fig. App. 3-7: Nonlinear regression for total Switzerland. Detailed results of the regression analysis

Tab. App. 3-1: Linear model for all species.

Land use type	Region	R^2	d.f.	F	Sigf-p	a	b	SE a	SE b
°agri_hi	c-alp	0.22	69	18.9	0.00 **	9.34	0.52	1.15	0.12
°agri_hi	jur	0.12	42	5.8	0.02 **	5.16	0.20	1.93	0.08
agri_hi	n-alp	0.04	108	4.6	0.04 **	9.68	-0.13	1.08	0.06
°agri_li	jur	0.30	42	17.9	0.00 **	15.37	-0.27	1.66	0.06
°agri_li	pla-4	0.30	37	15.9	0.00 **	24.28	-0.94	2.05	0.23
°artificial_hi	c-alp	0.33	69	33.3	0.00 **	6.20	3.15	1.38	0.54
°artificial_hi	jur	0.46	42	36.2	0.00 **	0.35	2.57	1.64	0.43
artificial_hi	n-alp	0.05	108	5.4	0.02 **	5.55	0.84	1.18	0.37
°artificial_hi	pla-4	0.29	37	14.8	0.00 **	8.15	1.58	2.61	0.41
°artificial_hi	s-alp	0.34	19	9.8	0.01 **	11.73	1.02	2.12	0.33
°artificial_li	c-alp	0.33	69	33.4	0.00 **	9.06	3.66	0.99	0.63
°artificial_li	jur	0.41	42	29.1	0.00 **	3.23	2.08	1.34	0.38
artificial_li	n-alp	0.07	108	7.6	0.01 **	6.06	0.99	0.91	0.36
°artificial_li	pla-4	0.25	37	12.6	0.00 **	11.52	1.22	1.96	0.34
°artificial_li	s-alp	0.39	19	11.9	0.00 **	11.71	1.09	1.98	0.31
°forest_hi	c-alp	0.16	69	12.9	0.00 **	10.78	0.73	1.02	0.20
°forest_hi	jur	0.22	42	11.5	0.00 **	3.23	0.85	1.97	0.25
forest_li	c-alp	0.08	69	6.2	0.02 **	22.70	-0.21	3.94	0.08
°forest_li	s-alp	0.36	19	10.8	0.00 **	33.38	-0.36	5.29	0.11
°agri_hi	s-alp	0.17	19	3.9	0.06 *	13.75	0.41	2.17	0.21
non-use	c-alp	0.05	69	4.0	0.05 *	17.13	-0.22	2.20	0.11
°non-use	s-alp	0.14	19	3.0	0.10 *	19.74	-0.46	2.44	0.27
agri_hi	pla-4	0.00	37	0.0	0.86	18.60	-0.02	7.22	0.13
agri_li	c-alp	0.01	69	0.4	0.53	14.21	-0.05	1.99	0.09
agri_li	n-alp	0.00	108	0.2	0.68	8.48	-0.02	1.70	0.06
agri_li	s-alp	0.00	19	0.0	0.94	16.79	-0.03	2.95	0.41
forest_hi	n-alp	0.01	108	1.1	0.30	7.11	0.18	0.94	0.18
forest_hi	pla-4	0.00	37	0.0	0.98	17.37	-0.01	3.21	0.36
forest_hi	s-alp	0.03	19	0.5	0.48	12.97	0.16	5.30	0.22
forest_li	jur	0.04	42	1.6	0.21	15.80	-0.16	5.29	0.13
forest_li	n-alp	0.00	108	0.4	0.51	6.08	0.04	2.69	0.07
forest_li	pla-4	0.01	37	0.2	0.65	19.60	-0.13	5.19	0.29

Land use type	Region	R^2	d.f.	F	Sigf-p	a	b	SE a	SE b
lakes	c-alp	0.00	69	0.1	0.72	12.71	0.16	1.33	0.45
lakes	jur	0.01	42	0.2	0.64	9.42	-0.37	1.03	0.77
lakes	n-alp	0.00	108	0.1	0.78	7.87	-0.02	0.68	0.07
lakes	pla-4	0.06	37	2.5	0.12	15.51	1.44	1.67	0.91
lakes	s-alp	0.04	19	0.7	0.41	14.05	1.18	3.50	1.41
non-use	jur	0.02	42	0.9	0.36	9.83	-1.65	1.15	1.77
non-use	n-alp	0.01	108	1.3	0.26	6.99	0.11	0.99	0.09
non-use	pla-4	0.00	37	0.1	0.78	17.09	0.72	1.49	2.54
agri_hi	Switzerl.	0.04	348	15.7	0.00 **	9.75	0.09		
agri_li	Switzerl.	0.12	348	47.9	0.00 **	16.26	-0.24		
artificial_hi	Switzerl.	0.18	348	78.5	0.00 **	7.32	1.10		
artificial_li	Switzerl.	0.19	348	79.3	0.00 **	8.51	1.03		
forest_hi	Switzerl.	0.06	348	22.4	0.00 **	9.64	0.34		
forest_li	Switzerl.	0.06	348	20.6	0.00 **	16.97	-0.14		
lakes	Switzerl.	0.00	348	0.0	0.88	11.97	-0.01		
non-use	Switzerl.	0.01	348	2.1	0.15	12.48	-0.08		

where R^2: regression coefficient, doff.: degree of freedom, F: F-statistics for testing the hypothesis that slope b = 0, Sigf-p: significance value, * $p<0.01$, ** $p<0.05$, a/b: parameter, SE a/SE b: standard error for the parameter. Regressions taken in the damage model into account (marked with "°") are those which are significant and show a $R^2 \geq 10\%$.

Tab. App. 3-2: Linear model for threatened species.

Land use type	Region	R^2	d.f.	F	Sigf-p	*a*	*b*	SE *a*	SE *b*
°agri_hi	jur	0.10	42	4.5	0.04 **	37.92	0.55	5.99	0.26
agri_hi	n-alp	0.09	108	10.5	0.00 **	51.38	-0.62	3.42	0.19
°agri_li	jur	0.25	42	14.3	0.00 **	66.47	-0.76	5.25	0.20
°agri_li	pla-4	0.25	37	12.4	0.00 **	78.44	-2.08	5.14	0.59
°artificial_hi	c-alp	0.23	69	20.1	0.00 **	33.64	5.67	3.21	1.26
°artificial_hi	jur	0.31	42	18.9	0.00 **	26.80	6.46	5.69	1.49
°artificial_hi	pla-4	0.19	37	8.6	0.01 **	44.90	3.12	6.75	1.06
°artificial_hi	s-alp	0.19	19	4.5	0.05 **	42.64	1.41	4.33	0.66
°artificial_li	c-alp	0.15	69	12.3	0.00 **	40.10	5.40	2.40	1.54
°artificial_li	jur	0.25	42	14.2	0.00 **	34.65	5.00	4.61	1.33
°artificial_li	pla-4	0.16	37	6.9	0.01 **	51.86	2.33	5.07	0.88
°artificial_li	s-alp	0.25	19	6.2	0.02 **	42.16	1.60	4.05	0.64
forest_hi	c-alp	0.06	69	4.0	0.05 **	43.08	0.94	2.33	0.47
°forest_hi	jur	0.21	42	11.4	0.00 **	30.78	2.61	6.04	0.77
forest_li	c-alp	0.06	69	4.3	0.04 **	63.54	-0.38	8.64	0.19
°forest_li	s-alp	0.23	19	5.8	0.03 **	74.18	-0.54	10.71	0.22
non-use	n-alp	0.04	108	4.6	0.03 **	37.48	0.64	3.13	0.30
artificial_li	n-alp	0.03	108	3.2	0.08 *	38.70	2.12	2.98	1.18
agri_hi	c-alp	0.03	69	2.0	0.16	43.10	0.41	2.76	0.29
agri_hi	pla-4	0.00	37	0.0	0.95	63.95	-0.02	17.53	0.33
agri_hi	s-alp	0.08	19	1.7	0.21	45.70	0.53	4.20	0.40
agri_li	c-alp	0.01	69	0.6	0.44	43.04	0.15	4.30	0.19
agri_li	n-alp	0.00	108	0.0	0.84	43.51	-0.04	5.51	0.19
agri_li	s-alp	0.00	19	0.1	0.79	48.20	0.20	5.44	0.75
artificial_hi	n-alp	0.02	108	1.8	0.19	38.21	1.59	3.85	1.20
forest_hi	n-alp	0.00	108	0.0	0.97	42.55	-0.02	3.06	0.57
forest_hi	pla-4	0.01	37	0.4	0.55	58.68	0.52	7.75	0.87
forest_hi	s-alp	0.01	19	0.1	0.76	46.42	0.13	9.89	0.41
forest_li	jur	0.01	42	0.6	0.44	61.73	-0.30	16.41	0.39
forest_li	n-alp	0.00	108	0.2	0.64	38.52	0.10	8.72	0.22
forest_li	pla-4	0.01	37	0.3	0.59	69.54	-0.38	12.59	0.69
lakes	c-alp	0.00	69	0.3	0.60	44.88	0.52	2.87	0.98

Land use type	Region	R^2	d.f.	F	Sigf-p	*a*	*b*	SE *a*	SE *b*
lakes	jur	0.03	42	1.5	0.23	50.55	-2.84	3.12	2.33
lakes	n-alp	0.00	108	0.2	0.63	42.74	-0.11	2.19	0.23
lakes	pla-4	0.03	37	1.3	0.26	59.75	2.56	4.12	2.25
lakes	s-alp	0.07	19	1.4	0.25	42.76	3.05	6.35	2.55
non-use	c-alp	0.03	69	2.1	0.15	52.58	-0.36	4.83	0.24
non-use	jur	0.02	42	0.8	0.37	50.92	-4.95	3.52	5.44
non-use	pla-4	0.01	37	0.3	0.59	63.98	-3.37	3.60	6.14
non-use	s-alp	0.05	19	0.9	0.34	52.78	-0.50	4.73	0.52
agri_hi	Switzerl.	0.02	348	8.1	0.01 **	44.44	0.16		
agri_li	Switzerl.	0.05	348	19.4	0.00 **	54.83	-0.38		
artificial_hi	Switzerl.	0.10	348	37.3	0.00 **	40.32	1.86		
artificial_li	Switzerl.	0.10	348	37.7	0.00 **	42.33	1.75		
forest_hi	Switzerl.	0.02	348	8.6	0.00 **	44.76	0.50		
forest_li	Switzerl.	0.03	348	12.4	0.00 **	57.38	-0.26		
lakes	Switzerl.	0.00	348	0.5	0.47	48.43	-0.15		
non-use	Switzerl.	0.01	348	3.1	0.08 *	49.69	-0.23		

Tab. App. 3-3: Exponential model for all species.

Land use type	Region	R^2	d.f.	F	Sigf-p	*a*	*b*	SE *a*	SE *b*
°agri_hi	c-alp	0.17	63	12.5	0.00 **	2.04	0.04	0.13	0.01
°agri_hi	jur	0.11	42	5.2	0.03 **	1.56	0.02	0.22	0.01
°agri_li	jur	0.31	42	18.5	0.00 **	2.69	-0.03	0.18	0.01
°agri_li	pla-4	0.43	37	28.0	0.00 **	3.57	-0.12	0.20	0.02
°artificial_hi	c-alp	0.24	69	21.5	0.00 **	1.84	0.25	0.14	0.05
°artificial_hi	jur	0.44	42	32.6	0.00 **	1.04	0.28	0.19	0.05
artificial_hi	n-alp	0.07	104	7.2	0.01 **	1.29	0.14	0.17	0.05
°artificial_hi	pla-4	0.19	37	8.5	0.01 **	1.82	0.14	0.31	0.05
°artificial_hi	s-alp	0.20	19	4.7	0.04 **	2.36	0.06	0.19	0.03
°artificial_li	c-alp	0.24	68	21.2	0.00 **	2.06	0.29	0.10	0.06
°artificial_li	jur	0.40	42	28.4	0.00 **	1.33	0.23	0.15	0.04
artificial_li	n-alp	0.07	104	7.9	0.01 **	1.41	0.14	0.13	0.05
°artificial_li	pla-4	0.15	37	6.7	0.01 **	2.14	0.10	0.23	0.04
°artificial_li	s-alp	0.24	19	6.1	0.02 **	2.35	0.07	0.18	0.03
forest_hi	c-alp	0.10	50	5.3	0.03 **	2.29	0.05	0.12	0.02
°forest_hi	jur	0.24	41	12.7	0.00 **	1.23	0.11	0.23	0.03
forest_li	c-alp	0.06	69	4.7	0.03 **	3.18	-0.02	0.37	0.01
°forest_li	s-alp	0.28	19	7.4	0.01 **	3.85	-0.03	0.45	0.01
non-use	c-alp	0.06	69	4.4	0.04 **	2.79	-0.02	0.20	0.01
agri_hi	n-alp	0.03	102	3.0	0.09 *	1.90	-0.02	0.16	0.01
°non-use	s-alp	0.16	19	3.7	0.07 *	2.94	-0.04	0.19	0.02
agri_hi	pla-4	0.03	37	1.3	0.26	1.75	0.02	0.78	0.01
agri_hi	s-alp	0.10	19	2.2	0.16	2.48	0.03	0.18	0.02
agri_li	c-alp	0.00	69	0.0	0.95	2.40	0.00	0.19	0.01
agri_li	n-alp	0.01	104	0.6	0.45	1.86	-0.01	0.26	0.01
agri_li	s-alp	0.01	19	0.2	0.69	2.58	0.01	0.24	0.03
forest_hi	n-alp	0.01	87	0.7	0.41	1.49	0.02	0.17	0.03
forest_hi	pla-4	0.01	37	0.5	0.51	2.86	-0.03	0.35	0.04
forest_hi	s-alp	0.07	19	1.3	0.26	2.21	0.02	0.42	0.02
forest_li	jur	0.02	42	0.9	0.35	2.55	-0.01	0.59	0.01
forest_li	n-alp	0.01	105	1.2	0.28	1.25	0.01	0.41	0.01
forest_li	pla-4	0.04	37	1.5	0.24	3.30	-0.04	0.56	0.03

Land use type	Region	R^2	d.f.	F	Sigf-p		*a*	*b*	SE *a*	SE *b*
lakes	c-alp	0.00	69	0.1	0.74		2.42	-0.01	0.12	0.04
lakes	jur	0.00	42	0.0	0.84		2.01	-0.02	0.12	0.09
lakes	n-alp	0.00	104	0.1	0.78		1.66	0.00	0.10	0.01
lakes	pla-4	0.06	37	2.3	0.14		2.45	0.15	0.19	0.10
lakes	s-alp	0.03	19	0.5	0.49		2.49	0.08	0.28	0.11
non-use	jur	0.01	42	0.5	0.50		2.05	-0.14	0.13	0.20
non-use	n-alp	0.01	105	1.4	0.23		1.56	0.02	0.14	0.01
non-use	pla-4	0.01	36	0.5	0.47		2.56	0.20	0.17	0.28
agri_hi	Switzerl.	0.02	335	8.3	0.00	**	1.98	0.01		
agri_li	Switzerl.	0.12	343	45.1	0.00	**	2.60	-0.02		
artificial_hi	Switzerl.	0.12	343	48.0	0.00	**	1.77	0.09		
artificial_li	Switzerl.	0.12	342	48.1	0.00	**	1.87	0.09		
forest_hi	Switzerl.	0.06	306	17.9	0.00	**	1.90	0.03		
forest_li	Switzerl.	0.02	344	8.4	0.00	**	2.50	-0.01		
lakes	Switzerl.	0.00	343	0.0	0.87		2.16	0.00		
non-use	Switzerl.	0.00	343	0.2	0.67		2.18	0.00		

Tab. App. 3-4: Exponential model for threatened species.

Land use type	Region	R^2	d.f.	F	Sigf-p	*a*	*b*	SE *a*	SE *b*
agri_hi	n-alp	0.08	102	8.7	0.00 **	3.84	-0.02	0.11	0.01
°agri_li	jur	0.23	42	12.6	0.00 **	4.25	-0.02	0.15	0.01
°agri_li	pla-4	0.34	37	19.1	0.00 **	4.54	-0.06	0.13	0.01
°artificial_hi	c-alp	0.20	69	17.1	0.00 **	3.46	0.14	0.08	0.03
°artificial_hi	jur	0.25	42	14.1	0.00 **	3.20	0.17	0.17	0.04
°artificial_hi	pla-4	0.14	37	6.2	0.02 **	3.64	0.07	0.18	0.03
°artificial_li	c-alp	0.12	68	9.4	0.00 **	3.62	0.13	0.06	0.04
°artificial_li	jur	0.21	42	11.4	0.00 **	3.40	0.13	0.14	0.04
°artificial_li	pla-4	0.12	37	4.9	0.03 **	3.81	0.05	0.14	0.02
°artificial_li	s-alp	0.20	19	4.8	0.04 **	3.69	0.03	0.10	0.02
°forest_hi	jur	0.19	41	9.4	0.00 **	3.24	0.07	0.19	0.02
forest_li	c-alp	0.06	69	4.0	0.05 **	4.20	-0.01	0.23	0.00
°forest_li	s-alp	0.21	19	5.2	0.04 **	4.42	-0.01	0.26	0.01
non-use	n-alp	0.05	105	5.9	0.02 **	3.41	0.02	0.10	0.01
agri_hi	jur	0.07	42	3.3	0.08 *	3.50	0.01	0.17	0.01
°artificial_hi	s-alp	0.15	19	3.5	0.08 *	3.70	0.03	0.11	0.02
artificial_li	n-alp	0.03	104	2.9	0.09 *	3.46	0.06	0.09	0.04
agri_hi	c-alp	0.03	63	2.1	0.15	3.66	0.01	0.08	0.01
agri_hi	pla-4	0.02	37	0.9	0.36	3.64	0.01	0.46	0.01
agri_hi	s-alp	0.07	19	1.4	0.25	3.77	0.01	0.10	0.01
agri_li	c-alp	0.01	69	0.6	0.43	3.68	0.00	0.11	0.00
agri_li	n-alp	0.00	104	0.1	0.79	3.54	0.00	0.18	0.01
agri_li	s-alp	0.01	19	0.2	0.65	3.80	0.01	0.13	0.02
artificial_hi	n-alp	0.02	104	2.4	0.12	3.42	0.06	0.12	0.04
forest_hi	c-alp	0.01	50	0.7	0.40	3.79	0.01	0.07	0.01
forest_hi	n-alp	0.00	87	0.1	0.74	3.60	-0.01	0.12	0.02
forest_hi	pla-4	0.00	37	0.0	1.00	4.06	0.00	0.21	0.02
forest_hi	s-alp	0.01	19	0.3	0.62	3.73	0.01	0.24	0.01
forest_li	jur	0.00	42	0.1	0.83	3.88	0.00	0.47	0.01
forest_li	n-alp	0.02	105	1.7	0.20	3.22	0.01	0.29	0.01
forest_li	pla-4	0.04	37	1.6	0.22	4.46	-0.02	0.33	0.02
lakes	c-alp	0.00	69	0.3	0.62	3.73	0.01	0.07	0.03

Land use type	Region	R^2	d.f.	F	Sigf-p		*a*	*b*	SE *a*	SE *b*
lakes	jur	0.02	42	0.8	0.39		3.81	-0.06	0.09	0.07
lakes	n-alp	0.00	104	0.0	0.92		3.58	0.00	0.07	0.01
lakes	pla-4	0.03	37	1.3	0.27		3.98	0.07	0.11	0.06
lakes	s-alp	0.06	19	1.2	0.29		3.70	0.07	0.15	0.06
non-use	c-alp	0.03	69	2.0	0.16		3.93	-0.01	0.13	0.01
non-use	jur	0.01	42	0.4	0.56		3.81	-0.09	0.10	0.16
non-use	pla-4	0.00	36	0.0	0.96		4.05	0.01	0.10	0.17
non-use	s-alp	0.06	19	1.3	0.28		3.94	-0.01	0.11	0.01
agri_hi	Switzerl.	0.01	335	4.0	0.05	**	3.68	0.00		
agri_li	Switzerl.	0.04	343	15.9	0.00	**	3.93	-0.01		
artificial_hi	Switzerl.	0.08	343	28.7	0.00	**	3.56	0.05		
artificial_li	Switzerl.	0.08	342	28.0	0.00	**	3.62	0.04		
forest_hi	Switzerl.	0.02	306	4.7	0.03	**	3.69	0.01		
forest_li	Switzerl.	0.01	344	5.0	0.03	**	3.92	0.00		
lakes	Switzerl.	0.00	343	0.1	0.77		3.76	0.00		
non-use	Switzerl.	0.00	343	0.4	0.53		3.77	0.00		

Appendix 4
Assignment of land use types

Tab. App. 4-1: Assignment of specific land use types to the 8 types of intensity of land use (number of investigated plots).

	agri_hi	agri_li	artificial_hi	artificial_li	forest_li	non-use	Total
Agricultural fallow						138	138
Agricultural fallow with hedgerows						11	11
Continuous urban			60				60
Conventional arable	54						54
Discontinuous urban			120				120
Fibre/energy crops (chinese reed)	50						50
Fibre/energy crops (hemp)	20						20
Fibre/energy crops (kenaf)	24						24
Forest edge						78	78
Green urban				110			110
Heathland						6	6
Hedgerows						44	44
Industrial area - part with vegetation			53				53
Industrial fallow						26	26
Integrated arable	157						157
Intensive meadow	69						69
Less intensive meadow		104					104
Mining fallow						10	10
Natural grassland		242					242
Organic arable	62						62
Organic meadow		727					727
Organic orchard		48					48
Peatbog						633	633
Rail embankments				57			57
Rail fallow						19	19

	agri_hi	agri_li	artificial_hi	artificial_li	forest_li	non-use	Total
Rural settlement				17			17
Semi-natural broad-leafed forest (arid)					223		223
Semi-natural broad-leafed forest (moist)					501		501
Sport facilities			16				16
Urban fallow						27	27
Totals	436	1121	249	184	724	992	3706

Tab. App. 4-2: Assignment of specific land use types to the Ecoinvent types (II, III, IV and III excluded), where the number of investigated plots is shown.

Land use type	II	III	IV	excluded III	Totals
Agricultural fallow	138				138
Agricultural fallow with hedgerows	11				11
Continuous urban			60		60
Conventional arable		54			54
Discontinuous urban			120		120
Fibre/energy crops (chinese reed)		50			50
Fibre/energy crops (hemp)		20			20
Fibre/energy crops (kenaf)		24			24
Forest edge	78				78
Green urban				110	110
Heathland	6				6
Hedgerows	44				44
Industrial area - part with vegetation			53		53
Industrial fallow	26				26
Integrated arable		157			157
Intensive meadow		69			69
Less intensive meadow		104			104
Mining fallow	10				10
Natural grassland	242				242
Organic arable		62			62
Organic meadow		727			727
Organic orchard		48			48
Peatbog	633				633
Rail embankments				57	57
Rail fallow	19				19
Rural settlement				17	17
Semi-natural broad-leafed forest (arid)	223				223
Semi-natural broad-leafed forest (moist)	501				501
Sport facilities			16		16

Land use type	II	III	IV	excluded III	Totals
Urban fallow	27				27
Totals	1958	1315	249	184	3706

Appendix 5a
Detailed results for local species richness considering all sources

Urban fabric (1.1)

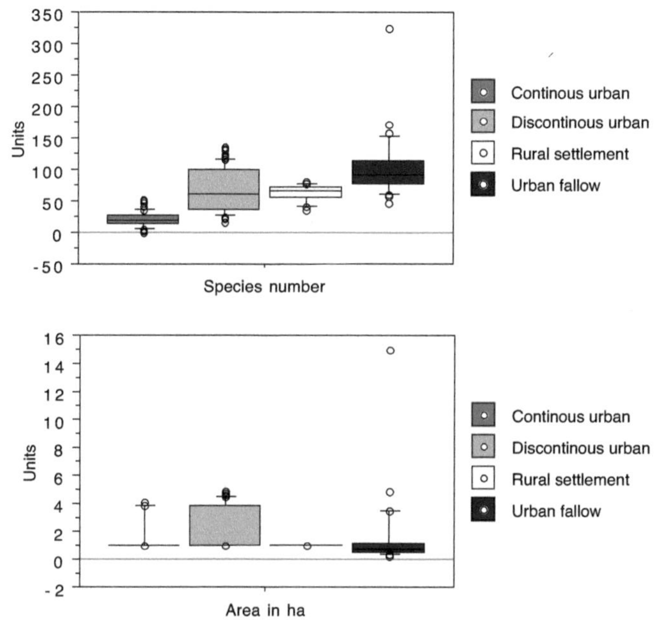

Fig. App. 5a-1

Tab. App. 5a-1

	Mean	Std. Dev.	Std. Error	Count	Minimum	Maximum	Median
Species number, Total	59.75	40.77	2.72	224	1.00	325.00	53.00
Species number, Continous urban	21.47	11.67	1.51	60	1.00	52.00	20.50
Species number, Discontinous urban	68.45	34.15	3.12	120	16.00	137.00	62.50
Species number, Rural settlement	63.41	13.21	3.20	17	36.00	80.00	68.00
Species number, Urban fallow	103.85	52.37	10.08	27	48.00	325.00	92.00
Area in ha, Total	1.98	1.68	.11	224	.20	15.00	1.00
Area in ha, Continous urban	1.46	1.05	.14	60	1.00	4.10	1.00
Area in ha, Discontinous urban	2.45	1.55	.14	120	1.00	4.90	1.00
Area in ha, Rural settlement	1.00	0.00	0.00	17	1.00	1.00	1.00
Area in ha, Urban fallow	1.66	2.90	.56	27	.20	15.00	.70

Industrial, commercial and transport (1.2)

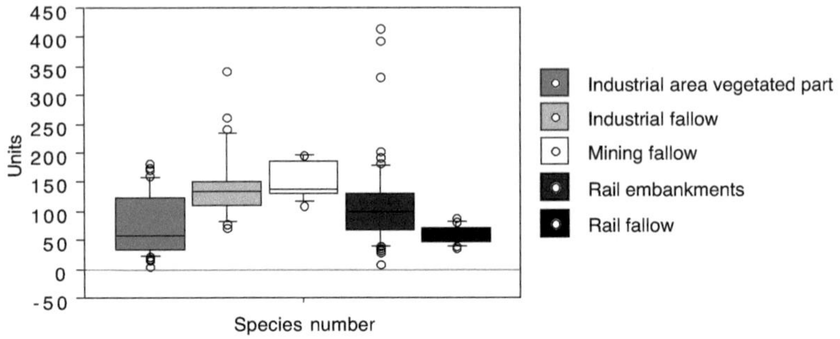

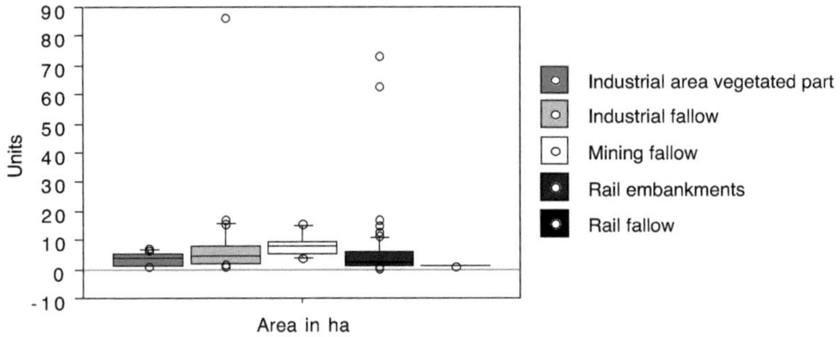

Fig. App. 5a-2

Tab. App. 5a-2

	Mean	Std. Dev.	Std. Error	Count	Minimum	Maximum	Median
Species number, Total	103.38	65.59	5.11	165	5.00	417.00	98.00
Species number, Industrial area vegetated part	81.75	52.55	7.22	53	5.00	183.00	58.00
Species number, Industrial fallow	144.12	59.62	11.69	26	72.00	344.00	135.50
Species number, Mining fallow	149.90	31.13	9.84	10	108.00	197.00	138.50
Species number, Rail embankments	110.88	77.19	10.22	57	9.00	417.00	98.00
Species number, Rail fallow	60.95	14.40	3.30	19	38.00	89.00	60.00
Area in ha, Total	5.15	10.18	.79	165	.60	86.40	2.40
Area in ha, Industrial area vegetated part	3.36	2.32	.32	53	1.00	7.40	4.00
Area in ha, Industrial fallow	8.37	16.51	3.24	26	1.20	86.40	4.40
Area in ha, Mining fallow	8.48	3.97	1.26	10	3.80	15.40	7.75
Area in ha, Rail embankments	6.14	12.55	1.66	57	.60	73.00	2.40
Area in ha, Rail fallow	1.00	0.00	0.00	19	1.00	1.00	1.00

Artificial non-agricultural areas with vegetation (1.4)

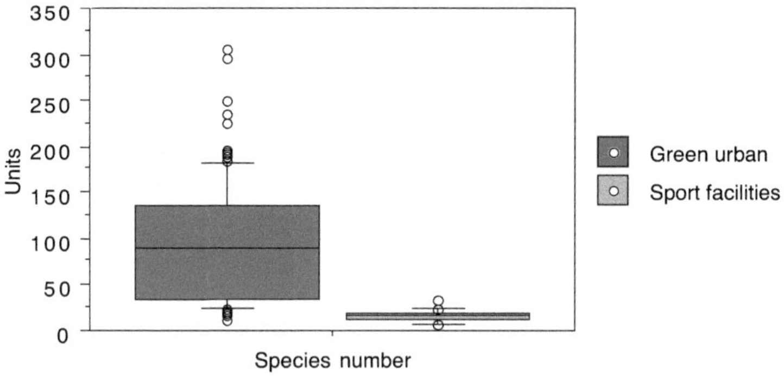

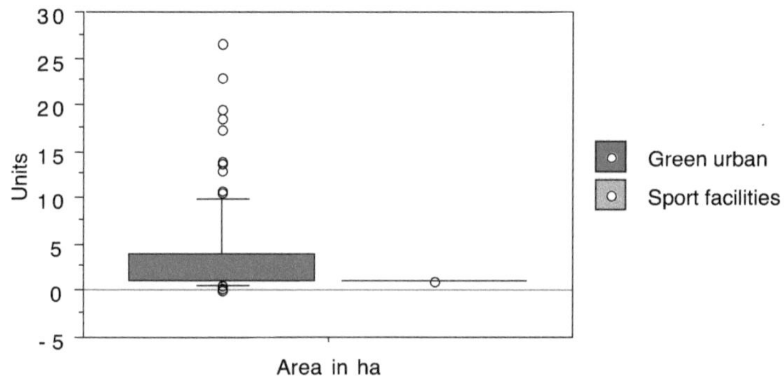

Fig. App. 5a-3

Tab. App. 5a-3

	Mean	Std. Dev.	Std. Error	Count	Minimum	Maximum	Median
Species number, Total	84.39	65.71	5.85	126	8.00	306.00	69.50
Species number, Green urban	94.15	64.71	6.17	110	13.00	306.00	91.00
Species number, Sport facilities	17.25	6.92	1.73	16	8.00	35.00	17.00
Area in ha, Total	3.26	4.68	.42	126	.20	26.50	1.00
Area in ha, Green urban	3.59	4.93	.47	110	.20	26.50	1.10
Area in ha, Sport facilities	1.00	0.00	0.00	16	1.00	1.00	1.00

Arable land (2.1)

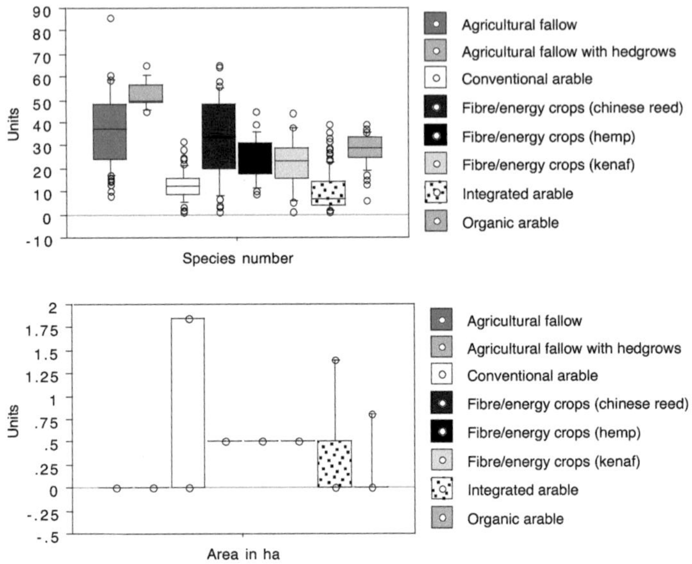

Fig. App. 5a-4

Tab. App. 5a-4

	Mean	Std. Dev.	Std. Error	Count	Minimum	Maximum	Median
Species number, Total	24.35	17.35	.76	516	1.00	86.00	21.00
Species number, Agricultural fallow	38.10	17.28	1.47	138	8.00	86.00	37.00
Species number, Agricultural fallow with hedgrows	52.91	5.99	1.81	11	45.00	65.00	50.00
Species number, Conventional arable	13.07	6.51	.89	54	1.00	32.00	12.00
Species number, Fibre/energy crops (chinese reed)	33.12	17.34	2.45	50	1.00	65.00	33.50
Species number, Fibre/energy crops (hemp)	23.75	9.37	2.09	20	9.00	45.00	22.50
Species number, Fibre/energy crops (kenaf)	21.96	11.04	2.25	24	1.00	44.00	23.00
Species number, Integrated arable	10.19	8.41	.67	157	1.00	39.00	7.00
Species number, Organic arable	28.44	6.89	.87	62	6.00	39.00	29.00
Area in ha, Total	.27	.43	.02	516	1.20E-3	1.85	.01
Area in ha, Agricultural fallow	.01	.01	4.37E-4	138	1.20E-3	.01	.01
Area in ha, Agricultural fallow with hedgrows	.01	0.00	0.00	11	.01	.01	.01
Area in ha, Conventional arable	.56	.85	.12	54	.01	1.85	.01
Area in ha, Fibre/energy crops (chinese reed)	.50	0.00	0.00	50	.50	.50	.50
Area in ha, Fibre/energy crops (hemp)	.50	0.00	0.00	20	.50	.50	.50
Area in ha, Fibre/energy crops (kenaf)	.50	0.00	0.00	24	.50	.50	.50
Area in ha, Integrated arable	.34	.45	.04	157	1.20E-3	1.40	.01
Area in ha, Organic arable	.16	.31	.04	62	.01	.80	.01

Pastures and meadows (2.3)

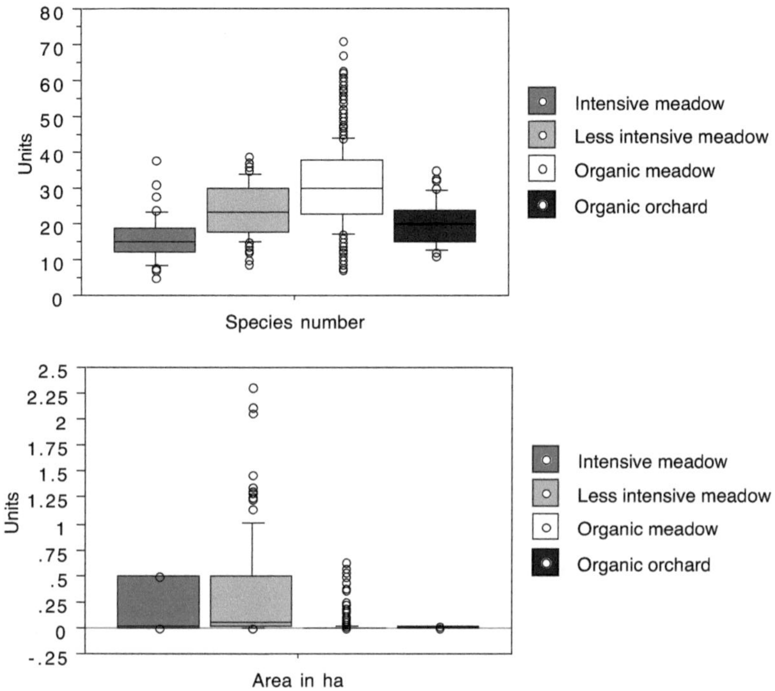

Fig. App. 5a-5

Tab. App. 5a-5

	Mean	Std. Dev.	Std. Error	Count	Minimum	Maximum	Median
Species number, Total	28.37	10.85	.35	948	5.00	71.00	28.00
Species number, Intensive meadow	15.90	6.20	.75	69	5.00	38.00	15.00
Species number, Less intensive meadow	24.00	7.47	.73	104	9.00	39.00	23.50
Species number, Organic meadow	30.71	10.59	.39	727	7.00	71.00	30.00
Species number, Organic orchard	20.27	6.18	.89	48	11.00	35.00	20.00
Area in ha, Total	.07	.22	.01	948	8.00E-4	2.30	1.20E-3
Area in ha, Intensive meadow	.14	.22	.03	69	1.20E-3	.50	.01
Area in ha, Less intensive meadow	.34	.49	.05	104	1.20E-3	2.30	.06
Area in ha, Organic meadow	.03	.10	3.80E-3	727	8.00E-4	.62	1.00E-3
Area in ha, Organic orchard	.01	4.19E-3	6.05E-4	48	1.20E-3	.01	.01

Forest (3.1)

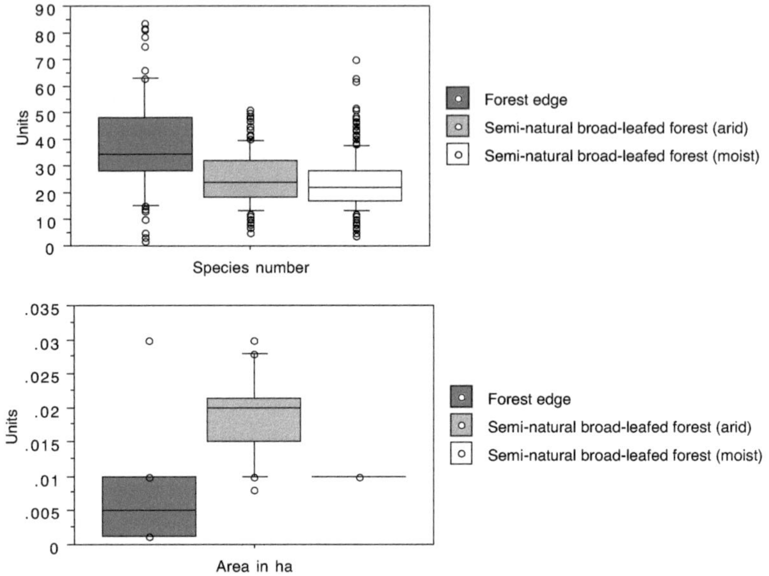

Fig. App. 5a-6

Tab. App. 5a-6

	Mean	Std. Dev.	Std. Error	Count	Minimum	Maximum	Median
Species number, Total	25.43	11.79	.42	802	2.00	84.00	23.00
Species number, Forest edge	37.91	18.37	2.08	78	2.00	84.00	34.50
Species number, Semi-natural broad-leafed forest (arid)	25.33	10.03	.67	223	5.00	51.00	24.00
Species number, Semi-natural broad-leafed forest (moist)	23.53	9.91	.44	501	4.00	70.00	22.00
Area in ha, Total	.01	.01	2.03E-4	802	1.20E-3	.03	.01
Area in ha, Forest edge	.01	.01	6.06E-4	78	1.20E-3	.03	.01
Area in ha, Semi-natural broad-leafed forest (arid)	.02	.01	3.97E-4	223	.01	.03	.02
Area in ha, Semi-natural broad-leafed forest (moist)	.01	0.00	0.00	501	.01	.01	.01

Shrub and/or herbaceous vegetation types (3.2)

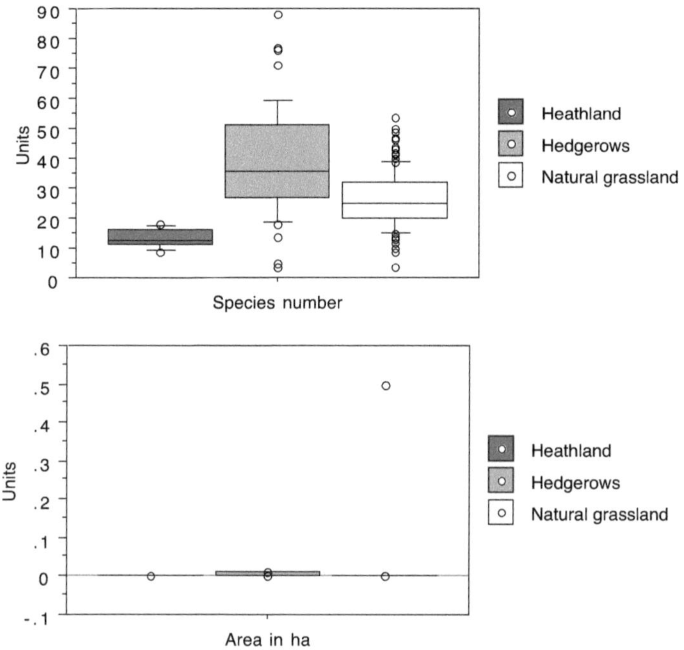

Fig. App. 5a-7

Tab. App. 5a-7

	Mean	Std. Dev.	Std. Error	Count	Minimum	Maximum	Median
Species number, Total	27.57	11.93	.70	292	4.00	88.00	26.50
Species number, Heathland	13.17	3.31	1.35	6	9.00	18.00	12.50
Species number, Hedgerows	38.41	18.41	2.78	44	4.00	88.00	35.50
Species number, Natural grassland	25.95	9.05	.58	242	4.00	54.00	25.00
Area in ha, Total	.01	.06	3.39E-3	292	8.00E-4	.50	1.60E-3
Area in ha, Heathland	2.50E-3	0.00	0.00	6	2.50E-3	2.50E-3	2.50E-3
Area in ha, Hedgerows	.01	3.96E-3	5.98E-4	44	1.20E-3	.01	.01
Area in ha, Natural grassland	.01	.06	4.09E-3	242	8.00E-4	.50	1.50E-3

Wetlands (4.1)

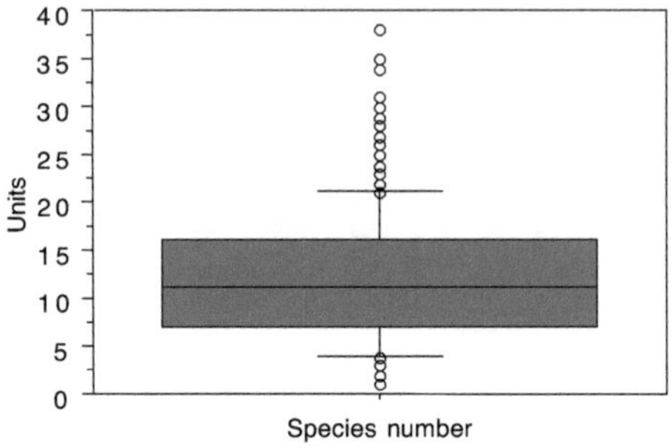

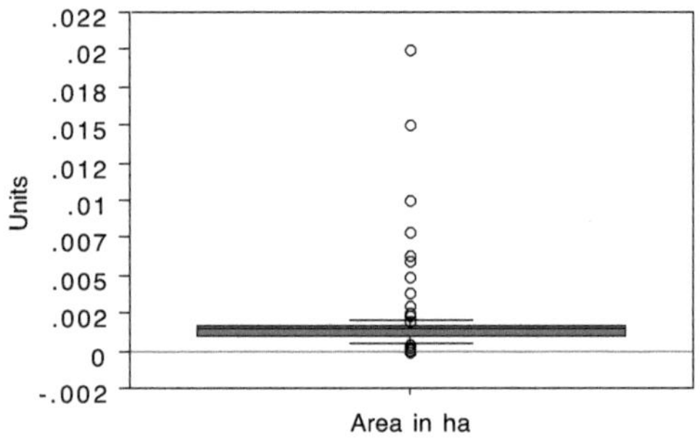

Fig. App. 5a-8

Tab. App. 5a-8

	Mean	Std. Dev.	Std. Error	Count	Minimum	Maximum	Median
Species number, Total	12.00	6.79	.27	633	1.00	38.00	11.00
Species number, Peatbog	12.00	6.79	.27	633	1.00	38.00	11.00
Area in ha, Total	1.78E-3	2.39E-3	9.50E-5	633	1.00E-5	.02	1.50E-3
Area in ha, Peatbog	1.78E-3	2.39E-3	9.50E-5	633	1.00E-5	.02	1.50E-3

Appendix 5b
Detailed results for local species richness considering specific data sources

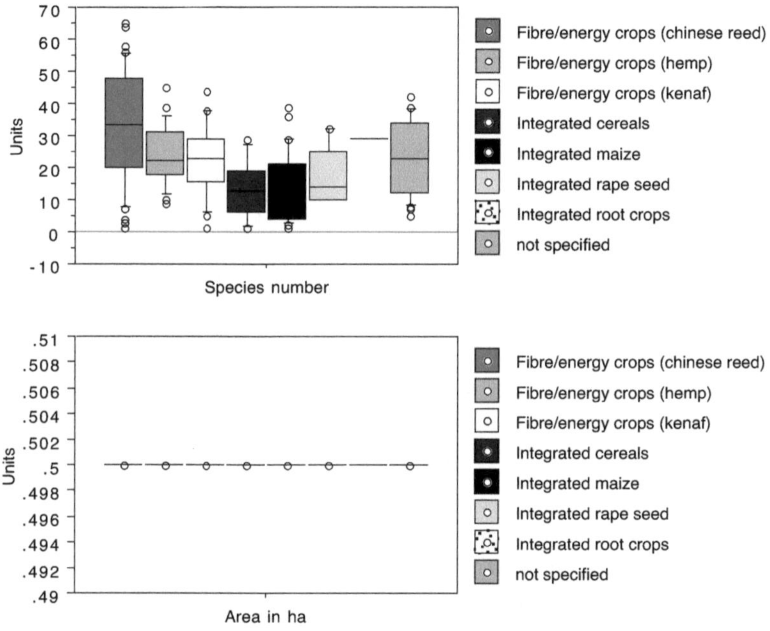

Fig. App. 5b-1: Agricultural fields, source Lips (1997), species number (top graph) and area in ha (bottom graph).

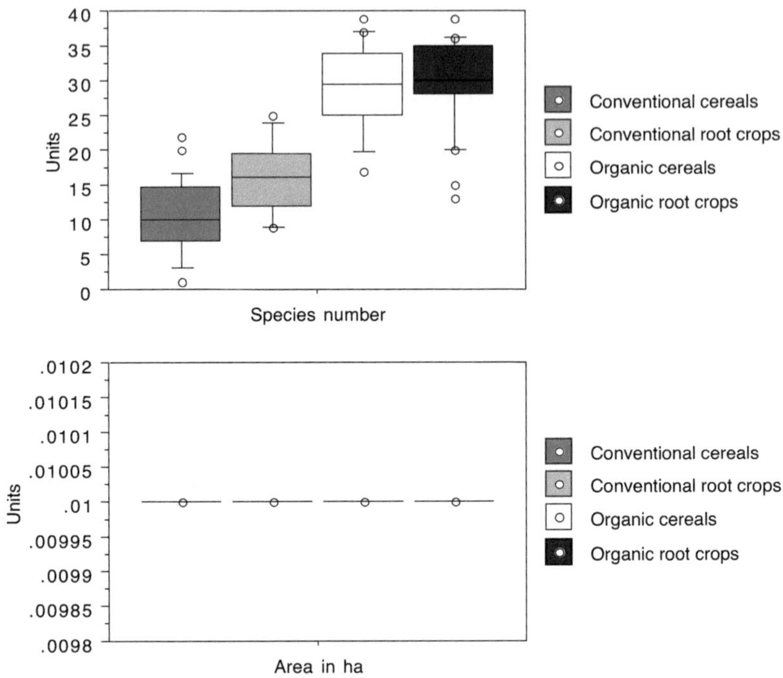

Fig. App. 5b-2: Agricultural fields, source Callauch (1981), species number (top graph) and area in ha (bottom graph).

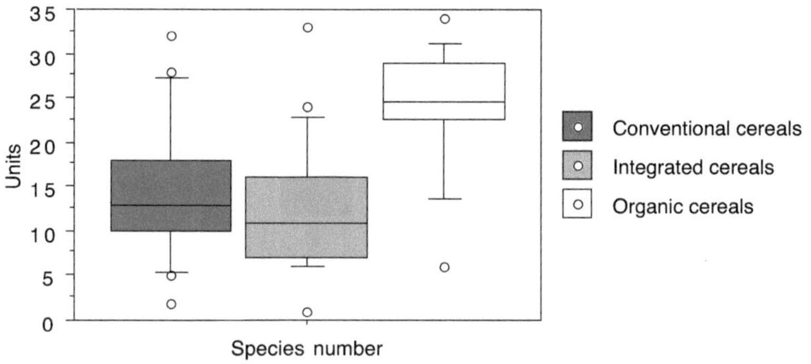

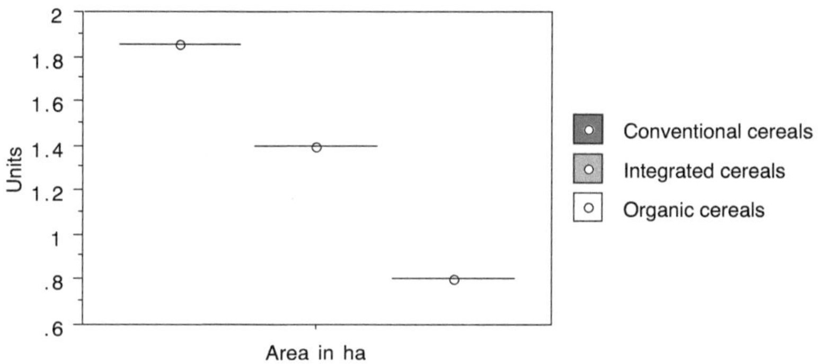

Fig. App. 5b-3: Agricultural fields, source Schreiber (1995), species number (top graph) and area in ha (bottom graph).

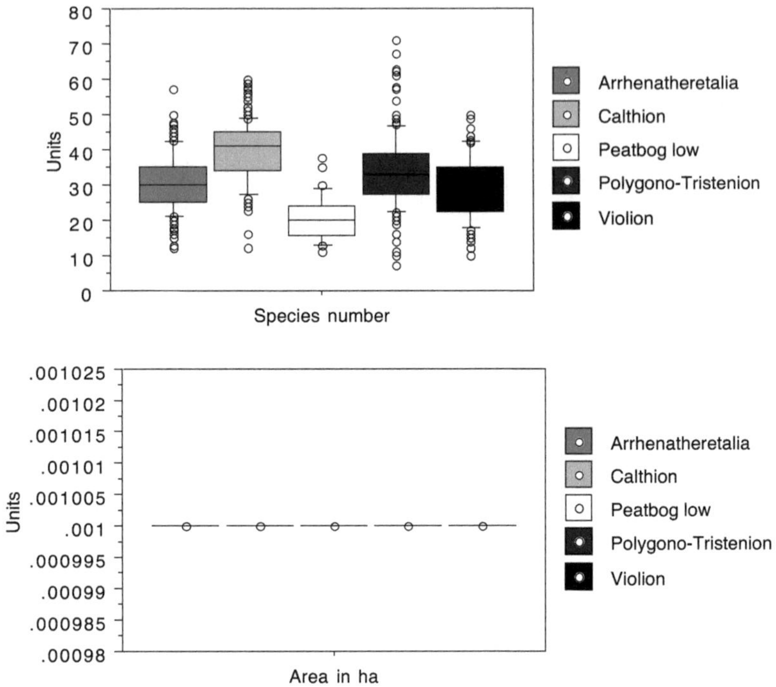

Fig. App. 5b-4: Different types of meadows, peatbogs and forests, source Bruelheide (1995), species number (top graph) and area in ha (bottom graph).

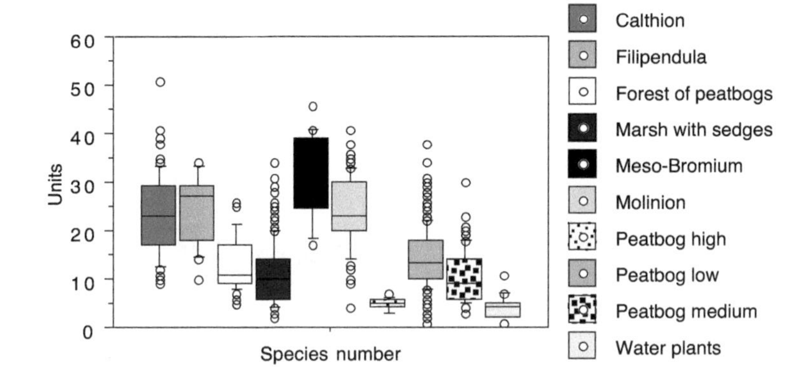

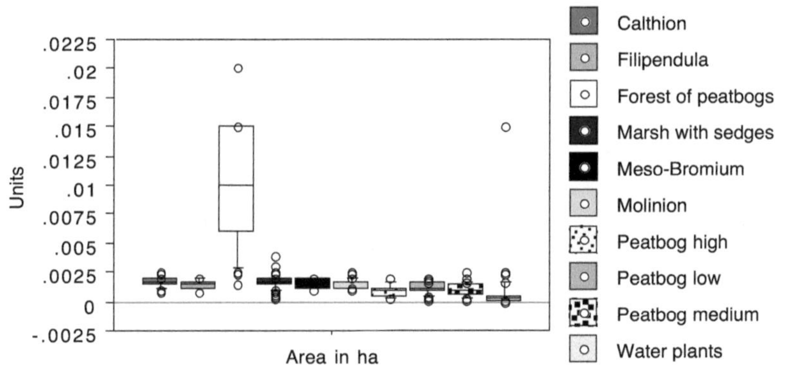

Fig. App. 5b-5: Different types of meadows, marshland and peatbogs, source Grüttner (1990), species number (top graph) and area in ha (bottom graph).

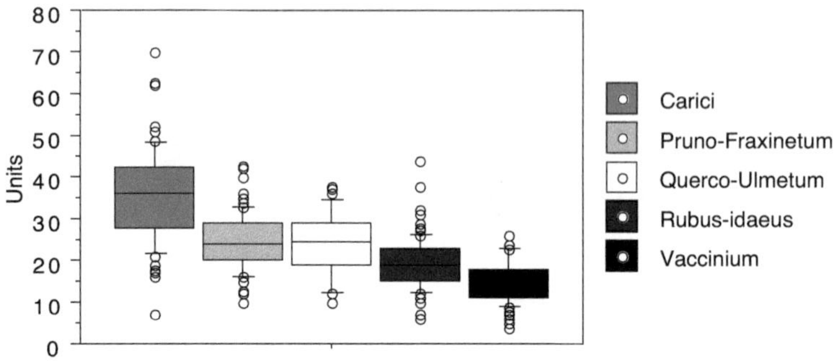

Species number

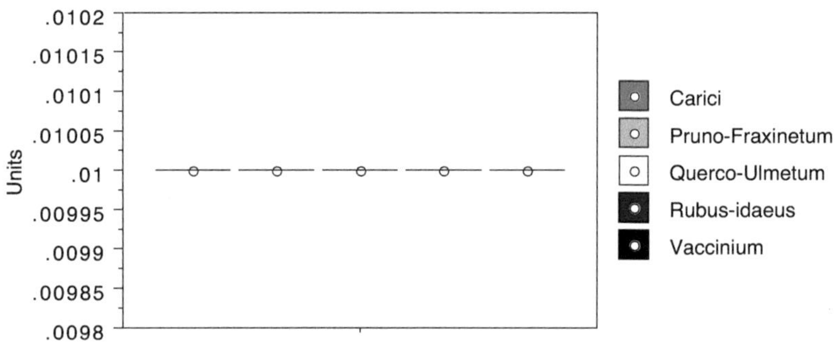

Area in ha

Fig. App. 5b-6: Comparison of forest types, source Döring-Mederake (1991), species number (top graph) and area in ha (bottom graph).

Appendix 6a
Overview with all characterisation factors $EDP^{sp\text{-}div}$

Tab. App. 6a-1: Local, regional and total Ecosystem damage potential (EDP) (next side).

		Local damage				Regional damage							Total damage							
							all *S*			threatened *S*			all *S*				threatened *S*			
		$EDP_{local-lin-pla}$	$EDP_{local-lin-c-alp}$	$EDP_{local-noni-pla}$	$EDP_{local-noni-c-alp}$	Type of intensity	$EDP^*_{region.-lin}$	$EDP^*_{region.-noni-pla}$	$EDP^*_{region.-noni-c-alp}$	$EDP^*_{region.-lin}$	$EDP^*_{region.-noni-pla}$	$EDP^*_{region.-noni-c-alp}$	$EDP_{total-lin-pla}$	$EDP_{total-lin-c-alp}$	$EDP_{total-noni-pla}$	$EDP_{total-noni-c-alp}$	$EDP_{total-lin-pla}$	$EDP_{total-lin-c-alp}$	$EDP_{total-noni-pla}$	$EDP_{total-noni-c-alp}$
CORINE+	*Land-use type*																			
2111	Conventional arable	0.74	0.79	0.36	0.43	agri_hi	0.09	0.18	0.05	–	–	–	0.41	0.44	0.27	0.24	–	–	–	–
2112	Integrated arable	0.82	0.86	0.47	0.53	agri_hi	0.09	0.18	0.05	–	–	–	0.45	0.47	0.32	0.29	–	–	–	–
2113	Organic arable	0.35	0.49	0.12	0.18	agri_hi	0.09	0.18	0.05	–	–	–	0.22	0.29	0.15	0.12	–	–	–	–
21141	Fibre/energ crops (kenaf)	0.75	0.80	0.38	0.44	agri_hi	0.09	0.18	0.05	–	–	–	0.42	0.45	0.28	0.25	–	–	–	–
21142	Fibre/energ crops (hemp)	0.73	0.79	0.36	0.42	agri_hi	0.09	0.18	0.05	–	–	–	0.41	0.44	0.27	0.24	–	–	–	–
21143	Fibre/energ crops (chinese reed)	0.63	0.71	0.27	0.33	agri_hi	0.09	0.18	0.05	–	–	–	0.36	0.40	0.22	0.19	–	–	–	–
2311	Intensive meadow	0.58	0.67	0.24	0.30	agri_hi	0.09	0.18	0.05	–	–	–	0.33	0.38	0.21	0.18	–	–	–	–
2212	Organic orchard	0.41	0.53	0.14	0.21	agri_li	-0.14	-0.20	-0.09	-0.32	-0.51	-0.33	0.14	0.20	-0.03	0.06	0.04	0.11	-0.19	-0.06
2312	Less intensive meadow	0.51	0.62	0.19	0.26	agri_li	-0.14	-0.20	-0.09	-0.32	-0.51	-0.33	0.19	0.24	0.00	0.08	0.10	0.15	-0.16	-0.03
2313	Organic meadow	-0.14	0.10	-0.04	0.03	agri_li	-0.14	-0.20	-0.09	-0.32	-0.51	-0.33	-0.14	-0.02	-0.12	-0.03	-0.23	-0.11	-0.27	-0.15
321	Natural grassland	0.02	0.23	0.01	0.07	agri_li	-0.14	-0.20	-0.09	-0.32	-0.51	-0.33	-0.06	0.05	-0.09	-0.01	-0.15	-0.05	-0.25	-0.13
10	Built up land	0.97	0.98	0.99	1.06	artificial_hi	0.47	0.91	0.38	0.94	1.59	0.93	0.72	0.73	0.95	0.72	0.96	0.96	1.29	1.00
111	Continuous urban	0.80	0.85	0.44	0.50	artificial_hi	0.47	0.91	0.38	0.94	1.59	0.93	0.64	0.66	0.68	0.44	0.87	0.89	1.01	0.72
112	Discontinuous urban	0.45	0.56	0.16	0.22	artificial_hi	0.47	0.91	0.38	0.94	1.59	0.93	0.46	0.52	0.54	0.30	0.69	0.75	0.87	0.58
1212	Industrial area - part with vegetation	0.39	0.52	0.13	0.20	artificial_hi	0.47	0.91	0.38	0.94	1.59	0.93	0.43	0.49	0.52	0.29	0.66	0.73	0.86	0.57
142	Sport facilities	0.83	0.87	0.48	0.54	artificial_hi	0.47	0.83	0.36	0.94	1.15	0.80	0.65	0.67	0.66	0.45	0.89	0.90	0.81	0.67
114	Rural settlement	0.38	0.51	0.13	0.19	artificial_li	0.45	0.83	0.36	0.81	1.15	0.80	0.42	0.48	0.48	0.27	0.59	0.66	0.64	0.49
1224	Rail embankments	0.20	0.37	0.06	0.13	artificial_li	0.45	0.83	0.36	0.81	1.15	0.80	0.33	0.41	0.45	0.24	0.51	0.59	0.60	0.46
141	Green urban	0.27	0.42	0.08	0.15	artificial_li	0.45	0.83	0.36	0.81	1.15	0.80	0.36	0.44	0.46	0.25	0.54	0.62	0.62	0.47

		Local damage				Regional damage							Total damage							
							all *S*			threatened *S*			all *S*				threatened *S*			
		$EDP_{local-lin-pla}$	$EDP_{local-lin-c-alp}$	$EDP_{local-nonl-pla}$	$EDP_{local-nonl-c-alp}$	Type of intensity	$EDP^*_{region.-lin}$	$EDP^*_{region.-nonl-pla}$	$EDP^*_{region.-nonl-c-alp}$	$EDP^*_{region.-lin}$	$EDP^*_{region.-nonl-pla}$	$EDP^*_{region.-nonl-c-alp}$	$EDP_{total-lin-pla}$	$EDP_{total-lin-c-alp}$	$EDP_{total-nonl-pla}$	$EDP_{total-nonl-c-alp}$	$EDP_{total-lin-pla}$	$EDP_{total-lin-c-alp}$	$EDP_{total-nonl-pla}$	$EDP_{total-nonl-c-alp}$
3112	Semi-natural broad-leafed forest (arid)	0.43	0.55	0.15	0.22	forest_li	-0.08	-0.19	-0.09	-0.12	-0.16	-0.12	0.17	0.23	-0.02	0.06	0.15	0.21	0.00	0.05
3112	Semi-natural broad-leafed forest (moist)	0.41	0.53	0.14	0.20	forest_li	-0.08	-0.19	-0.09	-0.12	-0.16	-0.12	0.16	0.22	-0.02	0.06	0.14	0.20	-0.01	0.04
113	Urban fallow	0.00	0.21	0.00	0.06	non-use	-0.10	-0.17	-0.09	–	–	–	-0.05	0.05	-0.08	-0.01	–	–	–	–
1225	Rail fallow	0.40	0.53	0.14	0.20	non-use	-0.10	-0.17	-0.09	–	–	–	0.15	0.21	-0.01	0.06	–	–	–	–
125	Industrial fallow	-0.01	0.20	0.00	0.06	non-use	-0.10	-0.17	-0.09	–	–	–	-0.06	0.05	-0.09	-0.01	–	–	–	–
134	Mining fallow	0.04	0.25	0.01	0.08	non-use	-0.10	-0.17	-0.09	–	–	–	-0.03	0.07	-0.08	-0.01	–	–	–	–
2115	Agricultural fallow	-0.09	0.14	-0.02	0.04	non-use	-0.10	-0.17	-0.09	–	–	–	-0.10	0.02	-0.10	-0.02	–	–	–	–
245	Agricultural fallow with hedgrows	-0.34	-0.05	-0.08	-0.01	non-use	-0.10	-0.17	-0.09	–	–	–	-0.22	-0.08	-0.12	-0.05	–	–	–	–
314	Forest edge	-0.20	0.05	-0.05	0.01	non-use	-0.10	-0.17	-0.09	–	–	–	-0.15	-0.03	-0.11	-0.04	–	–	–	–
322	Heathland	0.56	0.65	0.22	0.28	non-use	-0.10	-0.17	-0.09	–	–	–	0.23	0.27	0.03	0.10	–	–	–	–
325	Hedgerows	-0.12	0.11	-0.03	0.03	non-use	-0.10	-0.17	-0.09	–	–	–	-0.11	0.00	-0.10	-0.03	–	–	–	–
412	Peatbog	0.53	0.63	0.21	0.27	non-use	-0.10	-0.17	-0.09	–	–	–	0.21	0.26	0.02	0.09	–	–	–	–
Types of intensity	Agri_hi	0.68	0.75	0.31	0.37		0.09	0.18	0.05	–	–	–	0.38	0.42	0.24	0.21	–	–	–	–
	Agri_li	-0.02	0.19	-0.01	0.06		-0.14	-0.20	-0.09	-0.32	-0.51	-0.33	-0.08	0.03	-0.10	-0.02	-0.17	-0.06	-0.26	-0.13
	Artificial_hi	0.54	0.64	0.21	0.28		0.47	0.91	0.38	0.94	1.59	0.93	0.51	0.56	0.56	0.33	0.74	0.79	0.90	0.61
	Artificial_li	0.26	0.41	0.08	0.14		0.45	0.83	0.36	0.81	1.15	0.80	0.36	0.43	0.46	0.25	0.53	0.61	0.61	0.47
	Forest_hi	0.67	0.74	0.30	0.36		0.18	0.22	0.14	0.59	0.77	0.51	0.43	0.46	0.26	0.25	0.63	0.67	0.54	0.44
	Forest_li	0.41	0.54	0.14	0.21		-0.08	-0.19	-0.09	-0.12	-0.16	-0.12	0.17	0.23	-0.02	0.06	0.15	0.21	-0.01	0.04
	Non-use	0.31	0.46	0.10	0.17		-0.10	-0.17	-0.09	–	–	–	0.10	0.18	-0.03	0.04	–	–	–	–
Ecoinvent	II	0.31	0.46	0.10	0.17	non-use	-0.10	-0.17	-0.09	–	–	–	0.12	0.19	-0.03	0.04	–	–	–	–
	III	0.20	0.37	0.06	0.12	agri_hi	0.09	0.18	0.05	–	–	–	0.13	0.21	0.12	0.09	–	–	–	–
	IV	0.54	0.64	0.21	0.28	artificial_hi	0.47	0.91	0.38	0.94	1.59	0.93	0.47	0.51	0.56	0.33	0.75	0.79	0.90	0.61
	III excluded	0.26	0.41	0.08	0.14		–	–	–	–	–	–	–	–	–	–	–	–	–	–

EDP values are given for CORINE+land use types, for types of intensity and Ecoinvent types. For the regional scale the standardised EDP^* = EDP/4.41 is shown. For EDP_{total} local and regional EDP are weighted each with 50%. (S: species, lin: linear model, nonlin: nonlinear model, pla: Swiss plateau, c-alp: Central Alps);

Appendix 6b
Results of Monte-Carlo simulation for total $EDP^{sp\text{-}div}$ with all the species

Tab. App. 6b-1 Results of Monte-Carlo simulation for total $EDP^{sp\text{-}div}$ with all the species (next page)

	$EDP_{total-lin-pla}$					$EDP_{total-lin-c-alp}$					$EDP_{total-nonlin-pla}$					$EDP_{total-nonlin-c-alp}$				
	Mean	Std Dev.	25% Perc.	50% Perc.	75% Perc.	Mean	Std Dev.	25% Perc.	50% Perc.	75% Perc.	Mean	Std Dev.	25% Perc.	50% Perc.	75% Perc.	Mean	Std Dev.	25% Perc.	50% Perc.	75% Perc.
CORINE+ No.																				
2111	0.41	0.09	0.35	0.42	0.47	0.44	0.07	0.39	0.44	0.49	0.30	0.12	0.22	0.28	0.37	0.25	0.10	0.19	0.23	0.29
2112	0.45	0.07	0.40	0.45	0.50	0.47	0.06	0.44	0.47	0.51	0.35	0.13	0.25	0.32	0.41	0.30	0.11	0.22	0.28	0.34
2113	0.21	0.14	0.13	0.23	0.31	0.28	0.10	0.22	0.30	0.35	0.17	0.10	0.11	0.16	0.22	0.13	0.07	0.08	0.12	0.16
21141	0.41	0.07	0.37	0.42	0.46	0.45	0.06	0.41	0.45	0.48	0.31	0.11	0.23	0.29	0.36	0.26	0.09	0.20	0.25	0.30
21142	0.41	0.06	0.37	0.41	0.45	0.44	0.05	0.41	0.44	0.47	0.29	0.10	0.22	0.28	0.34	0.25	0.07	0.20	0.24	0.28
21143	0.35	0.11	0.28	0.36	0.43	0.40	0.08	0.34	0.40	0.45	0.25	0.12	0.17	0.23	0.31	0.21	0.10	0.15	0.19	0.24
2311	0.32	0.12	0.25	0.32	0.40	0.38	0.10	0.32	0.38	0.45	0.24	0.12	0.16	0.22	0.30	0.19	0.10	0.13	0.17	0.23
2212	0.12	0.11	0.06	0.13	0.20	0.19	0.09	0.14	0.20	0.25	-0.02	0.06	-0.06	-0.03	0.01	0.06	0.06	0.03	0.06	0.09
2312	0.18	0.15	0.09	0.19	0.27	0.24	0.12	0.16	0.24	0.32	0.01	0.10	-0.05	0.00	0.05	0.10	0.10	0.04	0.08	0.14
2313	-0.16	0.28	-0.34	-0.15	0.03	-0.04	0.23	-0.17	-0.01	0.12	-0.10	0.08	-0.15	-0.12	-0.07	-0.02	0.08	-0.07	-0.03	0.01
321	-0.08	0.22	-0.20	-0.06	0.06	0.03	0.17	-0.07	0.04	0.14	-0.09	0.08	-0.13	-0.10	-0.06	0.00	0.07	-0.04	-0.01	0.03
10	0.72	0.02	0.71	0.72	0.74	0.72	0.02	0.71	0.72	0.74	0.98	0.18	0.86	0.96	1.09	0.72	0.10	0.66	0.71	0.77
111	0.64	0.06	0.60	0.64	0.67	0.66	0.05	0.62	0.66	0.69	0.71	0.18	0.58	0.69	0.81	0.44	0.09	0.38	0.43	0.48
112	0.45	0.12	0.37	0.46	0.54	0.51	0.10	0.44	0.51	0.58	0.57	0.17	0.44	0.54	0.66	0.30	0.08	0.25	0.29	0.34
1212	0.42	0.18	0.31	0.44	0.54	0.49	0.14	0.39	0.49	0.58	0.56	0.18	0.43	0.53	0.65	0.29	0.11	0.22	0.27	0.34
142	0.65	0.04	0.62	0.65	0.68	0.67	0.04	0.64	0.67	0.69	0.69	0.14	0.58	0.67	0.77	0.45	0.08	0.40	0.44	0.49
114	0.40	0.09	0.35	0.41	0.46	0.48	0.07	0.44	0.49	0.53	0.50	0.13	0.40	0.48	0.58	0.27	0.05	0.23	0.27	0.30
1224	0.32	0.19	0.20	0.33	0.45	0.40	0.14	0.31	0.41	0.50	0.47	0.15	0.37	0.46	0.56	0.24	0.08	0.19	0.23	0.28
141	0.34	0.20	0.22	0.35	0.49	0.43	0.17	0.32	0.44	0.55	0.49	0.16	0.38	0.48	0.59	0.26	0.10	0.19	0.24	0.30
311	0.16	0.14	0.08	0.17	0.26	0.23	0.10	0.17	0.24	0.30	-0.02	0.10	-0.07	-0.02	0.03	0.08	0.09	0.03	0.06	0.12
3112	0.15	0.14	0.07	0.16	0.24	0.22	0.11	0.16	0.23	0.30	-0.02	0.09	-0.07	-0.03	0.03	0.08	0.09	0.02	0.06	0.11
113	-0.08	0.21	-0.20	-0.07	0.07	0.04	0.15	-0.06	0.05	0.15	-0.08	0.08	-0.13	-0.09	-0.03	0.00	0.06	-0.04	-0.01	0.03
1225	0.14	0.09	0.09	0.15	0.21	0.21	0.07	0.16	0.21	0.26	-0.01	0.06	-0.05	-0.01	0.03	0.07	0.04	0.04	0.06	0.09
125	-0.07	0.18	-0.18	-0.06	0.05	0.04	0.14	-0.04	0.05	0.14	-0.08	0.07	-0.13	-0.09	-0.04	0.00	0.05	-0.04	-0.01	0.02
134	-0.05	0.13	-0.12	-0.04	0.04	0.06	0.09	0.01	0.08	0.13	-0.08	0.05	-0.12	-0.08	-0.04	0.00	0.03	-0.02	0.00	0.02
2115	-0.11	0.22	-0.23	-0.08	0.03	0.01	0.17	-0.10	0.02	0.13	-0.09	0.07	-0.14	-0.10	-0.05	-0.01	0.06	-0.05	-0.02	0.02
245	-0.25	0.15	-0.33	-0.23	-0.15	-0.09	0.11	-0.15	-0.08	-0.01	-0.13	0.05	-0.16	-0.12	-0.09	-0.05	0.03	-0.06	-0.05	-0.03
314	-0.18	0.35	-0.39	-0.16	0.06	-0.04	0.27	-0.21	-0.02	0.14	-0.10	0.10	-0.16	-0.11	-0.05	-0.02	0.09	-0.07	-0.03	0.02
322	0.22	0.08	0.17	0.22	0.27	0.27	0.06	0.23	0.27	0.31	0.03	0.06	-0.02	0.03	0.07	0.11	0.04	0.08	0.10	0.13
325	-0.12	0.33	-0.32	-0.10	0.10	-0.01	0.25	-0.16	0.01	0.16	-0.08	0.11	-0.15	-0.10	-0.04	-0.01	0.10	-0.07	-0.02	0.03

	$EDP_{total-lin-pla}$					$EDP_{total-lin-c-alp}$					$EDP_{total-nonlin-pla}$					$EDP_{total-nonlin-c-alp}$				
	Mean	Std Dev.	25% Perc.	50% Perc.	75% Perc.	Mean	Std Dev.	25% Perc.	50% Perc.	75% Perc.	Mean	Std Dev.	25% Perc.	50% Perc.	75% Perc.	Mean	Std Dev.	25% Perc.	50% Perc.	75% Perc.
412	0.21	0.15	0.10	0.22	0.31	0.26	0.11	0.18	0.26	0.33	0.04	0.12	-0.03	0.02	0.09	0.11	0.09	0.04	0.09	0.15
Types of intensity																				
Agri_hi	0.38	0.13	0.30	0.38	0.47	0.42	0.10	0.35	0.42	0.48	0.27	0.13	0.18	0.25	0.33	0.22	0.13	0.14	0.19	0.26
Agri_li	-0.11	0.28	-0.29	-0.09	0.08	0.01	0.22	-0.12	0.02	0.17	-0.09	0.09	-0.14	-0.10	-0.05	0.00	0.10	-0.06	-0.02	0.04
Artificial_hi	0.50	0.15	0.40	0.50	0.60	0.55	0.12	0.48	0.56	0.63	0.60	0.18	0.47	0.58	0.70	0.33	0.12	0.26	0.31	0.38
Artificial_li	0.35	0.20	0.23	0.36	0.48	0.43	0.16	0.33	0.44	0.54	0.49	0.15	0.38	0.47	0.58	0.26	0.10	0.19	0.24	0.30
Forest_hi	0.42	0.12	0.34	0.43	0.51	0.46	0.10	0.39	0.46	0.52	0.28	0.12	0.20	0.26	0.34	0.27	0.12	0.19	0.24	0.32
Forest_li	0.15	0.15	0.06	0.16	0.26	0.22	0.11	0.16	0.23	0.30	-0.02	0.09	-0.07	-0.02	0.03	0.08	0.09	0.02	0.06	0.11
Non-use	0.10	0.25	-0.05	0.10	0.25	0.17	0.19	0.05	0.18	0.29	-0.02	0.11	-0.10	-0.04	0.03	0.06	0.10	-0.01	0.04	0.10
Ecoinvent																				
III excluded	0.00	0.00	0.00	0.00	0.00	0.00	0.00	0.00	0.00	0.00	0.05	0.08	0.00	0.04	0.09	0.09	0.09	0.03	0.07	0.12
II	0.11	0.22	-0.03	0.12	0.25	0.18	0.17	0.07	0.19	0.30	-0.02	0.11	-0.09	-0.04	0.03	0.06	0.11	0.00	0.04	0.10
III	0.12	0.30	-0.06	0.12	0.30	0.20	0.24	0.05	0.21	0.36	0.15	0.13	0.06	0.13	0.21	0.11	0.12	0.03	0.08	0.16
IV	0.46	0.18	0.34	0.47	0.58	0.51	0.16	0.41	0.51	0.62	0.59	0.19	0.46	0.57	0.70	0.33	0.11	0.26	0.31	0.38

Appendix 6c
Results of Monte-Carlo simulation for total *EDP*$^{sp\text{-}div}$ with threatened species only

Tab. App. 6c-1 Results of Monte-Carlo simulation for total EDP$^{sp\text{-}div}$ *with threatened species only (next page)*

	$EDP_{total-lin-pla}$					$EDP_{total-lin-c-alp}$					$EDP_{total-nonlin-pla}$					$EDP_{total-nonlin-c-alp}$				
	Mean	Std Dev.	25% Perc.	50% Perc.	75% Perc.	Mean	Std Dev.	25% Perc.	50% Perc.	75% Perc.	Mean	Std Dev.	25% Perc.	50% Perc.	75% Perc.	Mean	Std Dev.	25% Perc.	50% Perc.	75% Perc.
CORINE+ No.																				
2111	–	–	–	–	–	–	–	–	–	–	–	–	–	–	–	–	–	–	–	–
2112	–	–	–	–	–	–	–	–	–	–	–	–	–	–	–	–	–	–	–	–
2113	–	–	–	–	–	–	–	–	–	–	–	–	–	–	–	–	–	–	–	–
21141	–	–	–	–	–	–	–	–	–	–	–	–	–	–	–	–	–	–	–	–
21142	–	–	–	–	–	–	–	–	–	–	–	–	–	–	–	–	–	–	–	–
21143	–	–	–	–	–	–	–	–	–	–	–	–	–	–	–	–	–	–	–	–
2311	–	–	–	–	–	–	–	–	–	–	–	–	–	–	–	–	–	–	–	–
2212	0.04	0.12	-0.04	0.04	0.12	0.10	0.09	0.05	0.11	0.16	-0.19	0.06	-0.23	-0.19	-0.14	-0.05	0.06	-0.09	-0.06	-0.03
2312	0.09	0.15	-0.01	0.10	0.19	0.14	0.12	0.06	0.14	0.22	-0.14	0.11	-0.22	-0.16	-0.10	-0.02	0.10	-0.08	-0.04	0.03
2313	-0.26	0.29	-0.44	-0.25	-0.06	-0.13	0.24	-0.26	-0.11	0.03	-0.26	0.09	-0.32	-0.28	-0.22	-0.14	0.09	-0.19	-0.15	-0.10
321	-0.16	0.21	-0.30	-0.15	-0.02	-0.05	0.17	-0.15	-0.05	0.07	-0.25	0.08	-0.30	-0.26	-0.21	-0.11	0.08	-0.16	-0.13	-0.08
10	0.96	0.06	0.92	0.96	1.00	0.96	0.07	0.92	0.96	1.01	1.31	0.24	1.15	1.29	1.45	1.02	0.13	0.93	1.01	1.09
111	0.87	0.08	0.81	0.87	0.92	0.89	0.08	0.84	0.89	0.94	1.04	0.24	0.87	1.01	1.18	0.74	0.12	0.66	0.73	0.81
112	0.68	0.14	0.60	0.69	0.78	0.75	0.12	0.68	0.75	0.83	0.89	0.23	0.73	0.87	1.03	0.60	0.11	0.52	0.59	0.67
1212	0.65	0.19	0.53	0.66	0.78	0.73	0.15	0.62	0.74	0.83	0.88	0.24	0.70	0.86	1.03	0.59	0.14	0.50	0.58	0.67
142	0.89	0.07	0.84	0.89	0.94	0.91	0.07	0.85	0.91	0.96	0.90	0.21	0.76	0.89	1.03	0.69	0.12	0.61	0.68	0.76
114	0.58	0.11	0.52	0.59	0.66	0.65	0.10	0.59	0.66	0.72	0.72	0.20	0.57	0.71	0.85	0.50	0.10	0.43	0.50	0.56
1224	0.49	0.19	0.38	0.50	0.62	0.58	0.16	0.48	0.59	0.69	0.69	0.21	0.55	0.67	0.82	0.48	0.12	0.40	0.47	0.55
141	0.54	0.22	0.40	0.55	0.69	0.62	0.17	0.50	0.63	0.74	0.71	0.22	0.55	0.69	0.85	0.49	0.14	0.41	0.48	0.56
311	0.15	0.15	0.06	0.16	0.24	0.21	0.11	0.14	0.21	0.28	0.02	0.11	-0.05	0.01	0.08	0.09	0.11	0.02	0.07	0.13
3112	0.13	0.15	0.04	0.14	0.23	0.20	0.11	0.13	0.21	0.27	0.01	0.11	-0.06	0.00	0.07	0.08	0.10	0.01	0.06	0.12
113	–	–	–	–	–	–	–	–	–	–	–	–	–	–	–	–	–	–	–	–
1225	–	–	–	–	–	–	–	–	–	–	–	–	–	–	–	–	–	–	–	–
125	–	–	–	–	–	–	–	–	–	–	–	–	–	–	–	–	–	–	–	–
134	–	–	–	–	–	–	–	–	–	–	–	–	–	–	–	–	–	–	–	–
2115	–	–	–	–	–	–	–	–	–	–	–	–	–	–	–	–	–	–	–	–
245	–	–	–	–	–	–	–	–	–	–	–	–	–	–	–	–	–	–	–	–
314	–	–	–	–	–	–	–	–	–	–	–	–	–	–	–	–	–	–	–	–
322	–	–	–	–	–	–	–	–	–	–	–	–	–	–	–	–	–	–	–	–

	$EDP_{total\text{-}lin\text{-}pla}$					$EDP_{total\text{-}lin\text{-}c\text{-}alp}$					$EDP_{total\text{-}nonlin\text{-}pla}$					$EDP_{total\text{-}nonlin\text{-}c\text{-}alp}$				
	Mean	Std Dev.	25% Perc.	50% Perc.	75% Perc.	Mean	Std Dev.	25% Perc.	50% Perc.	75% Perc.	Mean	Std Dev.	25% Perc.	50% Perc.	75% Perc.	Mean	Std Dev.	25% Perc.	50% Perc.	75% Perc.
325	–	–	–	–	–	–	–	–	–	–	–	–	–	–	–	–	–	–	–	–
412	–	–	–	–	–	–	–	–	–	–	–	–	–	–	–	–	–	–	–	–
Types of intensity																				
Agri_hi	–	–	–	–	–	–	–	–	–	–	–	–	–	–	–	–	–	–	–	–
Agri_li	-0.20	0.28	-0.37	-0.18	-0.02	-0.08	0.21	-0.22	-0.07	0.07	-0.25	0.09	-0.31	-0.26	-0.20	-0.12	0.10	-0.18	-0.14	-0.08
Artificial_hi	0.73	0.17	0.62	0.74	0.84	0.78	0.14	0.69	0.79	0.87	0.92	0.24	0.75	0.89	1.07	0.63	0.14	0.53	0.61	0.70
Artificial_li	0.53	0.21	0.37	0.53	0.68	0.61	0.17	0.51	0.61	0.72	0.71	0.21	0.55	0.69	0.83	0.49	0.12	0.41	0.48	0.56
Forest_hi	0.63	0.15	0.53	0.63	0.73	0.67	0.13	0.58	0.67	0.76	0.58	0.23	0.42	0.55	0.69	0.46	0.15	0.36	0.45	0.55
Forest_li	0.14	0.14	0.05	0.15	0.23	0.20	0.12	0.13	0.20	0.28	0.01	0.11	-0.06	0.00	0.07	0.09	0.11	0.02	0.07	0.13
Non-use	–	–	–	–	–	–	–	–	–	–	–	–	–	–	–	–	–	–	–	–
EcoInvent																				
III excluded	–	–	–	–	–	–	–	–	–	–	–	–	–	–	–	–	–	–	–	–
II	0.11	0.21	-0.02	0.12	0.25	0.19	0.16	0.09	0.20	0.30	–	–	–	–	–	–	–	–	–	–
III	0.10	0.31	-0.09	0.12	0.31	0.18	0.24	0.04	0.19	0.35	–	–	–	–	–	–	–	–	–	–
IV	0.46	0.19	0.34	0.48	0.58	0.50	0.16	0.40	0.50	0.62	0.93	0.24	0.76	0.91	1.07	0.62	0.14	0.53	0.61	0.70

Appendix 7
Life Cycle Inventory for the case study

*Tab. App. 7-1: Surfacetime [m^2*yr] necessary for production (in different agricultural techniques), storage, transportation and packaging of 1 kg vegetable (Jungbluth 2000).*

Land use activity	Prod. (greenh.) (1)	Prod. (less int.) (2)	Prod. (org.) (3)	Storage (4)	Transp. (5)	Packag. (6)
Land occupation						
Arable land, organic			1.54E-01			
Arable land, less intensive		1.10E-01				
Meadow, organic			5.12E-02			
Meadow, less intensive		3.65E-02				
Built up land (Ecoinvent IV)	3.47E-02					
Land transformation						
Ecoinvent II-III	1.21E-02	4.42E-04	3.83E-04	2.7E-03	5.2E-03	1.2E-04
Ecoinvent II-IV	2.32E-03	1.38E-04	1.20E-04	1.1E-04	1.1E-02	7.5E-06
Ecoinvent III-IV	3.48E-03	2.16E-04	1.43E-04	7.5E-05	7.3E-03	4.8E-06
Others						
Marine benthos	3.96E-02	1.51E-03	1.02E-03	5.9E-04	4.5E-03	4.7E-06
Total	9.22E-02	1.49E-01	2.07E-01	3.5E-03	2.8E-02	1.4E-04

(1) Production in greenhouses

(2) Production outdoors in less intensive mode

(3) Production outdoors in organic mode

(4) Storage of fresh vegetable by retailer

(5) Transport from average Europe to Switzerland (ca. 800 km)

(6) Packaging in plastic foil

Glossary

Biodiversity It includes the diversity on the genetic, the species and the ecological (e.g., ecosystem processes) level (Heywood and Baste 1995).

Ecosystem 1) Heterogeneous pattern of forests, agricultural, mountain, urban and aquatic ecosystems on the regional scale, also referred to as landscape ecosystem (Leser 1997). *or*

2) A homogenous part of the earth with its biotic (species/gildes) and abiotic (water, soil, air) components. Ecosystems are analysed with respect to its species composition and biomass amount or, on a more sophisticated level, with respect to its species-species interactions (e.g., predator<-> prey), its species-environment interactions (e.g., nutrients->species) and its flows of energy, matter and information. Examples for ecosystems are mixed forests, meadows and lakes, but also forest plantations and cities are ecosystems. Basically it is a descriptive/analytical concept.

Ecosystem functions They denote overall characteristics of ongoing processes in the landscape ecosystem. E.g., production of biomass, regulation of climate, fixation of solar energy, generation of soil, and soil fertility, pollination of crops, mitigation of floods and droughts, purification of air and water, decompostion of waste (Daily 1997a). The recreational or scientific use of ecosystems can also be regarded as a function. (*synonym*: Life support functions)

Ecosystem quality Ecosystem quality is a normative concept and divides into three attributes with value to stakeholders: (i) biodiversity, (ii) ecosystem functions, and (iii) ecological resources. All attributes correlate and capture different aspects of one system, the ecosystem. (*synonym*: land quality)

Land 1) Land, as a physical entity is a three-dimensional segment of the earth at the interface between lithosphere and atmosphere. It is defined in terms of its topography and spatial nature. *or*

2) A broader integrative view also includes natural entities like mineral body, soil, water, air and biota. These components are organised in ecosystems.

(adapted from UNCED 1992b, p. 1).

Land occupation An activity in order to fulfil the desired economic purpose of this part of land; includes physical and chemical interventions maintaining the existing land cover type.

Land transformation Land transformation is an activity in order to transform an initial land cover type into an new land cover type with desired properties; in-

	cludes physical and chemical interventions (*synonyms*: land cover change, land use change).
Land use	Activity in order to get access to natural resources in a controlled way; it implies land occupation and transformation, but also the phase of abandonment is seen as a part of the land use activity.
Land use types	Land use types are the result of land use activities, e.g., forest, arable land, urban fabric (*synonym*: land cover type).
Natural resource	a) Matter belonging to the earth's mineral body (like fossil oil, fossil water and ore), b) land surface, c) energy (e.g., solar radiation, geothermic energy, nuclear energy) and d) a part of ecosystem with instrumental value to human beings; either (i) exctractive stocks of abiotic (surface water, soil) or biotic components (genes, wild populations of animals/plants, food/non-food biomass of cultivated silvi-/agricultural systems) or (ii) non-exctractive stocks like soil and water quality (Bergh 1996, pp. 62). The type d) is named ecological resources in this book.
Parameter	Numerical values used in mathematical functions to map systemic relations. They can be considered constant for a specific case.
Reference state	Benchmark to compare different subjects with either just descriptive or normative implication.
Regional species-pool	Set of species occurring in a reasonably large region.
Resource	Anything which can contribute to economic activity. This includes natural resources, human resources, capital goods or man-made means of production. (Black 1997, p. 402)

Key to acronyms and abbreviations

A	Area
a	Parameter in various equations
Agri_hi	High-intensity agriculture
Agri_li	Low-intensity agriculture
Artificial_hi	High-intensity artificial area
Artificial_li	Low-intensity artificial area
b	Parameter in various equations
c-alp	Central Alps
EDP	Ecosystem damage potential (it is also used as simplified form of $EDP^{sp\text{-}div}$)
*EDP**	Standardised ecosystem damage potential
$EDP_{local–lin–c\text{-}alp}$	Local ecosystem damage potential calculated with linear model and Central Alps as reference
$EDP_{local–lin–pla}$	Local ecosystem damage potential calculated with linear model and Swiss Plateau as reference
$EDP_{local–nonl–c\text{-}alp}$	Local ecosystem damage potential calculated with nonlinear model and Central Alps as reference
$EDP_{local–nonl–pla}$	Local ecosystem damage potential calculated with nonlinear model and Swiss Plateau as reference
$EDP_{region.–lin}$	Regional ecosystem damage potential calculated with linear model
$EDP_{region.–nonl\text{-}c\text{-}alp}$	Regional ecosystem damage potential calculated with nonlinear model of Central Alps
$EDP_{region.–nonl\text{-}pla}$	Regional ecosystem damage potential calculated with nonlinear model of Swiss Plateau
$EDP^{sp\text{-}div}$	Ecosystem damage potential with focus on species diversity (*EDP* is used as simplified form of $EDP^{sp\text{-}div}$)
$EDP_{total–lin\text{-}c\text{-}alp}$	Index including $EDP_{local–lin–c\text{-}alp}$ and $EDP_{region.–lin}$
$EDP_{total–lin\text{-}pla}$	Index including $EDP_{local–lin–pla}$ and $EDP_{region.–lin}$
$EDP_{total–nonl\text{-}c\text{-}alp}$	Index including $EDP_{local–nonl–c\text{-}alp}$ and $EDP_{region.–nonl\text{-}c\text{-}alp}$
$EDP_{total–nonl\text{-}pla}$	Index including $EDP_{local–nonl–pla}$ and $EDP_{region.–nonl\text{-}pla}$
EQ	Ecosystem quality

Forest_hi	High-intensity forestry
Forest_li	Low-intensity forestry
jur	Jura
L[%]	Proportion of land use type in %
LCA	Life Cycle Assessment
LCIA	Life Cycle Impact Assessment
n-alp	Northern Alps
pla	Swiss Plateau
s-alp	Southern Alps
SE	Standard error
S_{lost}	Species number potentially lost on the regional scale
S_{occ}	Species number of occupied plot
SPAR	Species-area relationship
S_{ref}	Species number of reference
S_{region}	Average species number of a region
$S_{standard}$	Standardised species number on the local scale

Key to names of plant species and vegetation types in Latin, English, and German

Latin	English	German	
Cannabis sativa L.	Hemp	Hanf	Agriculture
Hibiscus cannabinus L.	Kenaf	Kenaf	Agriculture
Miscanthus sp.	Chinese reed	Chinagras	Agriculture
–	Silver fir	Tannenwald	Forest
–	Sycamore-ash wood	Ahorn-Eschenmischwälder	Forest
–	Lime wood	Lindenwald	Forest
–	Spruce wood	Fichtenwald	Forest
–	Mountain pine wood	Bergföhrenwald	Forest
–	Chestnut wood	Kastanienwald	Forest
–	Larch-arolla pine wood	Lärchen-Arven-Föhren-Wald	Forest
–	Xerothermic broadleafed forest	Wärmeliebender Laubwald	Forest
Carici	Fen alder wood	Erlenbruchwald	Forest
Fagus sylvatica, L.	Beech	Buche	Forest
Pruno-Fraxinetum	Alder ash wood	Traubenkirschen-Erlen Eschenwald	Forest
Querco-Ulmetum	Oak elm wood	Eichen-Ulmenauwald	Forest
Rubus-idaeus-Alnus glutinosa	Alder wood	Himbeer-Erlenwald	Forest
Vaccinium uliginosum-Betula pubescens	Fen birch wood	Birkenbruchwald	Forest
–	Fertilised meadows in valleys	Talfettwiese	Grassland
–	Wet litter meadows	nasse Streumagerwiese	Grassland
Arrhenatheretalia	Oat grass meadow	Glatthaferwiese	Grassland
Calthion	Fertilised moist meadow	Gedüngte Feuchtwiese	Grassland
Caricion	Small-sedge meadow	Kleinseggenriede	Grassland

Filipendula	Tall-herb communities	Mädesüsshoch-staudenflur	Grassland
Magnocaricion	Tall-sedge swamp	Grossseggenried	Grassland
Meso-Bromium	Mesoxeric grassland	Halbtrockenrasen	Grassland
Molinion	Moor-grass meadow	Pfeifengrasswiese	Grassland
Polygono	Oat grass meadow	Goldhaferwiese	Grassland
Violion	Mat-grass pasture	Borstgrassrasen	Grassland
Xero-Bromium	Xeric grassland	Trockenrasen	Grassland
e.g., Caricetum	Fen	Niedermoor	Moor
e.g., Scheuchze-rietum	Intermediate mires	Zwischenmoor	Moor
Sphagnetum	Raised bog	Hochmoor	Moor
Calluna	Dwarf-shrub heaths	Zwergstrauchheiden	other
Phragmition	Vegetation of filling up area (e.g., reed swamp)	Verlandungsgesell-schaften (e.g., Röhrichte)	other

Translation according to Ellenberg (1988, English edition) and Ellenberg (1996, German edition).

Literature

Albracht, R. (1997). *Zur Variabilität des Arteninventares verschiedener Bereiche von Fussballrasen, Golfplätzen und Mähweiden*. PhD thesis, Gießen, Fachbereich Agrarwissenschaften und Umweltsicherung, University Gießen.

Anonymus (1998). *LCA Graphic Paper and Print Products. Part 1: Proposal for a New Forestry Assessment Method in LCA*. Hamburg. Axel Springer.

Arrhenius, O. (1921). Species and area. *Journal of Ecology* 9: 95-99.

Attfield, R. (1998). Existence value and intrinsic value. *Ecological Economics* 24: 163-168.

Audsley, E., Ed. (1996). *Harmonisation of Environmental Life Cycle Assessment for Agriculture*. Silsoe, UK, European Comission DG VI Agriculture. Final Report Concerted Action AIR3-CT94-2028.

Backhaus, K., B. Erichson, W. Plinke and R. Weiber (1996). *Multivariate Analysemethoden. Eine anwendungsorientierte Einführung*, Springer.

Baitz, M., J. Kreißig and C. Schöch (1998). *Methode der Integration der Naturrauminanspruchnahme in Ökobilanzen*. Stuttgart, unpublished. IKP Ganzheitliche Bilanzierung, University Stuttgart.

Baitz, M., J. Kreißig and M. Wolf (2000). Methode zur Integration der Naturraum-Inanspruchnahme in Ökobilanzen. *Forstwirtschaftliches Centralblatt* 119: 128-149.

Barbier, E.B. (1995). Tropical wetland values and environmental functions. In: *Biodiversity Conservation: Problems and Policies*. Perrings, C.A., K.G. Mäler, C. Folke, C.S. Holling and B.-O. Jansson, Dordrecht, Kluwer: 147-169.

Barthlott, W. (1998). The uneven distribution of global biodiversity: A challenge for industrial and developing countries. In: *German Global Change Research 1998*. Ehlers, E. and T. Krafft, Bonn, German National Committee on Global Change Research: 36.

Bastian, O. and K.-F. Schreiber, Eds. (1999). *Analyse und ökologische Bewertung der Landschaft*. Heidelberg, Berlin, Spektrum Akademischer Verlag.

Baur, B., K.C. Ewald, B. Freyer and A. Erhardt (1997). *Ökologischer Ausgleich und Biodiversität*. Basel, Boston, Berlin, Birkhäuser.

Bergh, J.C.J.M., van den (1996). *Ecological Economics and Sustainable Development: Theory, Methods and Application*. Cheltenham, Elgar.

BFS (1992). *GEOSTAT. Arealstatistik der Schweiz 1979/85*. Bern, Bundesamt für Statistik, Sektor Raumnutzung (BFS).

BFS and BUWAL (1998). *Die Bodennutzung der Schweiz im europäischen Kontext. Integration der Arealstatistik in CORINE land cover*. Neuchâtel, Bundesamt für Statistik (BFS), Bundesamt für Umwelt, Wald und Landschaft (BUWAL).

Bigler, F., P. Jeanneret, A. Lips, B. Schüpbach, M. Waldburger and P. Fried (1998). Wirkungskontrolle der Öko-Massnahmen: Biologische Vielfalt. *Agrarforschung* 5: 379-382.

Bisby, F.A. (1995). Characterization of biodiversity. In: *Global Biodiversity Assessment*. Heywood, V.H. and R.T. Watson, New York, Cambridge University Press: 21-106.

Black, J. (1997). *A Dictionary of Economics*. Oxford, Oxford Univ. Press.

Blonk, H., E. Lindeijer and J. Broers (1997). Towards a methodology for taking physical degradation of ecosystems into account in LCA. *Int. J. LCA* 2(2): 91-98.

Blume, H.P. and H. Sukopp (1976). Ökologische Bedeutung anthropogener Bodenveränderung. *Schriftenreihe Vegetationskunde* 10: 75-89.

Booth, W.C., G.G. Colomb and J.M. Williams (1995). *The craft of research*. Chicago, London, The University of Chicago Press.

Bortz, J. (1993). *Statistik für Sozialwissenschaftler*. Berlin, Hamburg, New York, Springer.

Bossel, H. (1994). *Modeling and Simulation*. Braunschweig, Wiesbaden, Vieweg.

Braunschweig, A., R. Förster, P. Hofstetter and R. Müller-Wenk (1996). *Developments in LCA Valuation*. St. Gallen, IWÖ-Diskussionsbeitrag 42. Institute for Economy and the Environement (IWÖ), University St. Gallen.

Braunschweig, A. and R. Müller-Wenk (1993). *Ökobilanzen für Unternehmungen: eine Wegleitung für die Praxis*. Bern, Haupt.

Breckling, B. and F. Müller (1997). Der Ökosystembegriff aus heutiger Sicht. Grundstrukturen und Grundfunktionen von Ökosystemen. In: *Handbuch der Umweltforschung. Grundlagen und Anwendungen der Ökosystemforschung*. Fränzle, O., F. Müller and W. Schröder, Landsberg a. Lech, Ecomed: II-2.2, 1-21.

Bruelheide, H. (1995). *Die Grünlandgesellschaften des Harzes und ihre Standortbedingungen. Mit einem Beitrag zum Gliederungsprinzip auf der Basis von statistisch ermittelten Artengruppen*. Berlin, Stuttgart, Cramer.

Bundesumweltministerium and Umweltbundesamt, Eds. (1995). *Handbuch Umweltcontrolling*. München, Vahlen.

Bürgi, M. (1997). *Waldentwicklung im 19. und 20. Jahrhundert. Veränderungen in der Nutzung und Bewirtschaftung des Waldes und seiner Eigenschaften als Habitat am Beispiel der öffentlichen Waldungen im Zürcher Unter- und Weinland*. PhD thesis, Zurich, ETH.

Bürgi, M. (1999). A case study of forest change in the Swiss lowlands. *Landscape Ecology* 14: 567-575.

Caley, M.J. (1997). Local endemism and the relationship between local and regional diversity. *OIKOS* 79(3): 612-615.

Callauch, R. (1981). Ackerunkraut-Gesellschaften auf biologischen und konventionellen Äckern in der weiteren Umgebung von Göttingen. *Tuexenia. Mitteilungen der Floristisch-soziologischen Arbeitsgemeinschaft* 1: 25-37.

Cederberg, C. (1998). *Life Cycle Assessment of Milk Production. A Comparison of Conventional and Organic Farming*. Gothenburg, report no. 643. SIK.

Connor, E.F. and E.D. McCoy (1979). The statistics and biology of the species-area relationship. *The American Naturalist* 113(6): 791-833.

Consoli, F., D. Allen, I. Boustead, J. Fava, W. Franklin, A.A. Jensen, N. de Oude, R. Parrish, R. Perriman, D. Postlethwaite, J. Séguin and B. Vigon (1993). *Guidelines of Life-Cycle Assessment: A'Code of Practice'*. Brussels, SETAC Workshop held at Sesimbra, Portugal, 31 March-3 April 1993. Society for Environmental Chemistry and Toxicology (SETAC).

Costanza, R. (1989). What is ecological economics? *Ecological Economics* 1: 1-7.

Costanza, R., Ed. (1991). *Ecological economics: The science and management of sustainability*. New York, Columbia Univ. Press.

Costanza, R., L. Wainger, C. Folke and K.G. Mäler (1993). Modeling complex ecological economic systems. Toward an evolutionary, dynamic understanding of people and nature. *BioScience* 8: 545-555.

Council of Europe, UNEP and European Centre for Nature Conservation (1996). *The Pan-European Biological and Landscape Diversity Strategy. A Vision for Europe's Natural Heritage*. Strasburg, Geneva, Tilburg.

Cowell, S. (1998). *Environmental Life Cycle Assessment of Agricultural Systems: Integration into Decision-Making*. PhD thesis, Surrey, Centre of Environmental Strategy, University Surrey.

Daily, G.C. (1997a). Introduction: What are ecosystem functions? In: *Nature's Services. Societal Dependence on Natural Ecosystems*. Daily, G.C., Washington, D.C., Covelo, Island: 1-10.

Daily, G.C., Ed. (1997b). *Nature's Services. Societal Dependence on Natural Ecosystems*. Washington, D.C., Covelo, California, Island Press.

de Groot, R.S. (1992). *Functions of Nature. Evaluation of Nature in Environmental Planning, Management and Decision Making*. Amsterdam, Wolters-Noordhoff.

di Castri, F. and T. Younès (1996). Introduction: Biodiversity, the emergence of a new scientific field – Its perspectives and constraints. In: *Biodiveristy, Science and Development. Towards a New Partnership*. di Castri, F. and T. Younès, Oxon, UK, CAB International: 1-10.

Diemer, M. et al. (1997). An experimental protocol to assess the effects of plant diversity on ecosystem functioning utilized in a European research network. *Bulletin of the Geobotanical Institute ETH* 63: 95-107.

Döring-Mederake, U. (1991). *Feuchtwälder im nordwestdeutschen Tiefland. Gliederung – Ökologie – Schutz*. Göttingen, Erich Goltze.

Duelli, P. (1997). Biodiversity evaluation in agricultural landscapes: An approach at two different scales. *Agriculture, Ecosystems and Environment* 62: 81-91.

Duelli, P. and M.K. Obrist (1998). In search of the best correlates for local organismal biodiversity in cultivated areas. *Biodiversity and Conservation* 7(3): 297-309.

Ehrlich, P.R. and A.H. Ehrlich (1981). *Extinction. The causes and consequences of the disappearance of species*. New York, Random House.

Ekvall, T. (1997). *Land Use in Life Cycle Assessment. Pilot Study*. Gothenburg, unpublished paper presented at the COST E9 Workshop in Brussels 4.4.98. Chalmers Industriteknik, Chalmers University.

El Titi, A. (1991). The Lautenbach Project 1978-89: integrated wheat production on a commercial arable farm, south-west Germany. In: *The Ecology of Temperate Cereal Fields*. Firbank, L.G., N. Carter, J.F. Darbyshire and G.R. Potts, Oxford, Blackwell: 399-412.

El Titi, A. and H. Landes (1990). Integrated farming system of Lautenbach: A practical contribution toward sustainable agriculture in Europe. In: *Sustainable Agriculture Systems*. Edwards, C.A. et al., Delray Beach, Florida, St. Lucie Press: 265-286.

Ellenberg, H. (1985). Veränderung der Flora Mitteleuropas unter dem Einfluss von Düngung und Immissionen. *Schweiz. Z. Forstwes*. 136(1): 19-39.

Ellenberg, H. (1988). *Vegetation Ecology of Central Europe*. Cambridge, Cambridge University Press.

Ellenberg, H. (1989). Eutrophierung – das gravierendste Problem im Naturschutz? *Norddeutsche Naturschutzakademie-Berichte* 1(2): 1-70.

Ellenberg, H. (1996). *Vegetation Mitteleuropas mit den Alpen in ökologischer, dynamischer und historischer Sicht*. Stuttgart, Ulmer.

Ellenberg, H., R. Mayer and J. Schauermann, Eds. (1986). *Ökosystemforschung. Ergebnisse des Sollingprojektes 1966-1986*. Stuttgart, Ulmer.

European Environment Agency (1998). *Europe's Environment: The Second Assessment*. Copenhagen, report no. 0-08-043202-6. Office for Official Publications of the European Communities. Elsevier Science: 293.

European Environmental Agency (2000). *CORINE Land Cover*. Luxembourg, available at http://etc.satellus.se, Date 8.6.00. European Environmental Agency.

Finlay, B.J., et al. (1997). Microbial diversity and ecosystem function. *Oikos* 80: 209-213.

Finnveden, G. (1996). "Resources" and related impact categories. In: *Towards a Methodology for Life Cycle Impact Assessment*. Udo de Haes, H., Brüssel, Society of Environmental Chemistry and Toxicology (SETAC).

Flückiger, P.E. (1999). *Der Beitrag von Waldrandstrukturen zur regionalen Biodiversität*. Olten, Dissertation, University Basel.

Forman, R.T.T. (1995). *Land Mosaics. The Ecology of Landscapes and Regions*. Cambridge, Cambridge Univ Press.

Frieben, B. (1998). *Untersuchungen zur Förderung Arten- und Biotopschutzgerechter Nutzung und ökologischen Strukturvielfalt im Ökologischen Landbau*. Bonn, report no. 60. Landwirtschaftliche Fakultät, Rheinische Friedrich-Wilhelms-Universität Bonn.

Frischknecht, R., P. Hofstetter, I. Knoepfel, R. Dones and E. Zollinger (1996). *Ökoinventare für Energiesysteme. Grundlagen für den ökologischen Vergleich von Energiesystemen und den Einbezug von Energiesystemen in Ökobilanzen für die Schweiz. Band 1-3.* Zürich, Villigen/ Würenlingen, Gruppe Energie-Stoffe-Umwelt (ESU-ETHZ), Sektion Ganzheitliche Systemanalyse (Paul Scherrer Institut).

Fukarek, F. (1980). Über die Gefährdung der Flora der Nordbezirke der DDR. *Phytocoenologia* 7: 174-182.

Funtowicz, S.O. and J.R. Ravetz (1990). *Uncertainty and quality in science for policy.* Dordrecht, Kluwer.

Giegrich, J. and K. Sturm (1996). *Operationalisierung der Wirkungskategorie Naturraumbeanspruchung.* Heidelberg. Institut für Energie und Umwelt (IFEU).

Gilbert, O.L. (1994). *Städtische Ökosysteme.* Radebeul, Neumann.

Giller, K.E., et al. (1997). Agricultural intensification, soil biodiversity and agroecosystem function. *Applied Soil Ecology* 6: 3-16.

Goedkoop, M. (1995). *The Eco-indicator 95. Weighting Method for Environmental Effects that Damage Ecosystems or Human Health on an European Scale. Final Report.* Utrecht, Bilthoven, report no. 9523. NOH.

Goedkoop, M., P. Hofstetter, R. Müller-Wenk and R. Spriensma (1998). The Eco-Indicator 98 explained. *Int. J. LCA* 3(6): 352-360.

Goedkoop, M. and R. Spriensma (1999). *The Eco-Indicator 99. A Damage Oriented Method for Life Cycle Impact Assessment. Methodology Report.* Den Haag, available at http://www.pre.nl/eco-indicator99/ei99-reports.htm. Ministerie van Volkshuisvesting.

Grant, W.E. and P.B. Thompson (1997). Integrated ecological models: Simulation of socio-cultural constraints on ecological dynamics. *Ecological Modelling* 100: 43-59.

Grüttner, A. (1990). *Die Pflanzengesellschaften und Vegetationskomplxe der Moore des westlichen Bodenseegebietes.* Berlin, Stuttgart, Cramer.

Guinée, J. (1995). *Development of a methodology for the environemntal life-cycle assessment of products.* PhD thesis, Leiden, CML, Leiden University.

Häberli, R., C. Wyss, B. Praplan and C. Lüscher (1992). *Kulturboden-BODEN KULTUR: Vorschläge für eine haushälterische Nutzung des Bodens in der Schweiz.* Zürich, vdf.

Hampicke, U. (1991). *Naturschutz-Ökonomie.* Stuttgart, Ulmer.

Hanssen, O.J., T. Rydberg and A. Rønning (1996). Integrating Life-Cycle Assessments in product development and management. In: *Environmental Life-Cycle Assessment.* Curran, A.M., New York et al., McGraw-Hill.

Hawksworth, D.L. and M.T. Kalin-Arroyo (1995). Magnitude and distribution of biodiversity. In: *Global Biodiversity Assessment.* Heywood, V.H. and R.T. Watson, New York, Cambridge University Press: 107-192.

Heck, K.L., G. van Belle and D. Simberloff (1975). Explicit calculation of the rarefaction diversity measurement and the determination of sufficient sample size. *Ecology* 56(6): 1459-1461.

Hector, A., B. Schmid, C. Beierkuhnlein, M.C. Caldeira, M. Diemer, P.G. Dimitrakopoulos, J.A. Finn, H. Freitas, P.S. Giller, J. Good, R. Harris, P. Högberg, K. Huss-Danell, J. Joshi, A. Jumpponen, C. Körner, P.W. Leadley, M. Loreau, A. Minns, C.P.H. Mulder, G. O'Donovan, S.J. Otway, J.S. Pereira, A. Prinz, D.J. Read, M. Scherer-Lorenzen, E.-D. Schulze, D. Siamantziouras, E.M. Spehn, A.C. Terry, A.J. Troumbis, F.I. Woodward, S. Yachi and H.J. Lawton (1999). Plant diversity an productivity experiments in European grasslands. *Science* 286: 1123-1127.

Hegg, O., C. Béguin and H. Zoller (1993). *Atlas schutzwürdiger Vegetationstypen der Schweiz*. Bern, Bundesamt für Umwelt, Wald und Landschaft.

Heijungs, R., Ed. (1992). *Environmental Life Cycle Assessment of Products. II: Backgrounds*. Leiden, Centrum voor Milieukunde.

Heijungs, R. (1998). Towards eco-efficiency with LCA's prevention principle: an epistemological foundation of LCA using axioms. In: *Product Innovation and Eco-efficiency*. Klostermann, J.E.M. and A. Tukker, Boston, Dordrecht, London, Kluwer.

Heijungs, R., J. Guinée and G. Huppes (1997). *Impact Categories for Natural Resources and Land Use*. Leiden, report no. 138. Centre of Environmental Science (CML).

Heijungs, R. and P. Hofstetter (1996). Definitions of terms and symbols. In: *Towards a Methodology for Life Cycle Impact Assessment*. Udo de Haes, H., Brüssel, Society of Environmental Chemistry and Toxicology (SETAC): 39-48.

Held, M. and K. Kümmerer (1997). Preserving soils for life. Proposal for a 'Convention of sustainable use of soils'. *Gaia* 6(3): 212-216.

Heywood, V.H. and I. Baste (1995). Introduction. In: *Global Biodiversity Assessment*. Heywood, V.H. and R.T. Watson, New York, Cambridge University Press: 1-20.

Hoffmann, F. (1997). Grundlagen der ökosystemaren Modellbildung. Deterministische Modellierung. In: *Handbuch der Umweltforschung. Grundlagen und Anwendungen der Okosystemforschung*. Fränzle, O., F. Müller and W. Schröder, Landsberg a. Lech, Ecomed: V-2.1 1-14.

Hofstetter, P. (1994). Analyse und Vergleich von Energiesystemen. In: *Ökoinventare zur Beurteilung von Energiesystemen*. Anonymus, Bern, Bundesamt für Energiewirtschaft. Report-No. 30187.

Hofstetter, P. (1998). *Perspectives in Life Cycle Impact Assessment. A structured Approach to Combine Models of the Technosphere, Ecosphere and Valuesphere*. Boston, Dordrecht, London, Kluwer.

Holton-Andersen, J., N. Christensen, L.W. Walsted Kristiansen, P. Kristensen and L. Emborg, Eds. (1998). *The State of Environment in Denmark*. Copenhagen,

Roskilde, Ministry of Environment and Energy, National Environmental Research Institute. NERI Technical Report, unpublished.

Hooper, D.U. and P.M. Vitousek (1997). The effects of plant composition and diversity on ecosystem processes. *Science* 277: 1302-1305.

Hulbert, S.H. (1971). The nonconcept of species diversity: A critique and alternative parameters. *Ecology* 52(4): 577-586.

ISO (1997). *Environmental Management. Life Cycle Assessment. Principles and Framework*. Genève, ISO 14040:1997. International Organization for Standardization (ISO).

Jaeger, J. and M. Scheringer (1998). Transdisziplinarität: Problemorientierung ohne Methodenzwang. *Gaia* 7(1): 10-25.

Jantsch, E. (1989). System, Systemtheorie. In: *Handlexikon zur Wissenschaftstheorie*. Seiffert, H. and G. Radnitzky, München, dtv Wissenschaft: 329-338.

Janzen, H. (1997). Erscheinungsformen und Trends des Lebenszyklusdenkens aus der Perspektive umweltorientierter Unternehmensplanung. *Zeitschrift für angewandte Umweltforschung* 10(3): 313-326.

Jungbluth, N. (2000). *Umweltfolgen des Nahrungsmittelkonsums: Beurteilung von Produktmerkmalen auf Grundlage einer modularen Ökobilanz*. PhD thesis ETH Nr. 13499 (available at www.dissertation.de), Zurich, UNS, ETH.

Kaule, G. (1991). *Arten- und Biotopschutz*. Stuttgart, Ulmer.

Knoepfel, I. (1995). *Indikatorensystem für die ökologische Bewertung des Transportes von Energie*. Zürich, report no. 20. Laboratorium für Energiesysteme ETH Zürich.

Köllner, T. (2000). Species-pool effect potentials (SPEP) as a yardstick to evaluate land-use impacts on biodiversity. *Journal of Cleaner Production* 8: 293-311.

Korneck, D. and H. Sukopp (1988). *Rote Listen der in der Bundesrepublik Deutschland ausgestorbenen, verschollenen und gefährdeten Farn- und Blütenpflanzen und ihre Auswertung für den Arten- und Biotopschutz*. Bonn-Bad Godesberg, Schr.Reihe Vegetationskde. Heft 19.

Kortman, J.G.M., E.W. Lindeijer, H. Sas and M. Sprengers (1994). *Towards Single Indicator for Emission. An Exercise in Aggregating Environmental Effects*. Amsterdam. Milieukunde. Universiteit van Amsterdam.

Kreikebaum, H. (1995). Die Notwendigkeit ökologischer Informationen. *Umweltwirtschaftsforum* 3(1): 6-9.

Kromrey, H. (1998). *Empirische Sozialforschung. Modelle und Methoden der Datenerhebung und Datenauswertung*. Opladen, UTB für Wissenschaft.

Kytzia, S. (1995). *Die Ökobilanz als Bestandteil des betrieblichen Informationsmanagments*. PhD thesis, St. Gallen, University St. Gallen.

Lawton, J.H. (1996). The role of species in ecosystems: aspects of ecological complexity and biological diversity. In: *Biodiversity. An ecological perspecitve*. Abe, T., S.A. Levin and M. Higashi, New York, Springer.

Leser, H. (1997). *Landschaftsökologie. Ansätze, Modelle, Methodiken, Anwendung.* Stuttgart, Ulmer.

Lindeijer, E., R. Müller-Wenk, B. Steen, M. Baitz, J. Broers, C. Cederberg, G. Finweden, M. ten Houten, T. Köllner, B. Mattson, J. May, I. Renner, L. Mila i Canals, I. Renner and B. Weidema (2001). *Impact Assessment of Resources and Land Use.* Lindeijer, E., R. Müller-Wenk and B. Steen (editors and responsible authors), Brussels, SETAC, WIA2, Taskforce on resources and land: 32.

Lindeijer, E., M. van Kampen, P. Fraanje, H. van Dobben, G.J. Nabuurs, E. Schouwenberg, D. Prins and N. Dankers (1998). *Biodiversity and Life Support Indicators for Land Use Impacts in LCA.* Wageningen, Texel, Publication series raw materials No. 1998/07. IVAM ER, IBN-DLO.

Lindfors, L.-G., K. Christiansen, L. Hoffmann, Y. Virtanen, V. Juntilla, A. Leskinen, O.-L. Hanssen, A. Rønning, T. Ekvall and G. Finnveden (1995). *LCA-Nordic. Impact Assessment.* Copenhagen, Technical Report No. 10, TemaNord 1995:503. Nordic Council of Ministers.

Lips, A., D. Dubois and et al. (1997). Belebte Umwelt. In: *Beurteilung nachwachsender Rohstoffe in der Schweiz in den Jahren 1993 – 1996. Vergleichende Betrachtung von Produkten aus ausgewählten NWR und entsprechenden konventionellen Produkten bezüglich ihrer Umweltwirkungen und Wirtschaftlichkeit.* Wolfensberger, U. and F. Dinkel, Tänikon, Basel, FAT Carbotech.

Lubchenco, L., et al. (1991). The sustainability biosphere initiative: An ecological research agenda. *Ecology* 72(2): 371-412.

Lynch, T. and D. Wells (1998). Non-anthropocentrism? A killing objection. *Environmental Values* 7: 151-163.

Magurran, A.E. (1996). *Ecological Diversity and its Measurement.* London, Chapman & Hall.

Manz, E. (1997). Vegetation ehemals militärisch genutzter Übungsplätze und Flugplätze und deren Bedeutung für den Naturschutz. *Tuexenia* 17: 173-192.

Martinez-Alier, J., G. Munda and J. O'Neill (1999). Commensurability and compensability in ecological economics. In: *Valuation and the Environment.* O'Connor, M. and S. Clive, Cheltenham, Northhampton, Edwaed Elgar.

May, R.M., J.H. Lawton and N.E. Stork (1995). Assessing extinction rates. In: *Extinction Rates.* Lawton, J.H. and R.M. May, Oxford, New York, Tokjo, Oxford Univ Press: 1-34.

McCloskey, J.M. and H.A. Spalding (1989). A reconnaissance-level inventory of the amount of wilderness remaining in the world. *Ambio* 18: 221-227.

McGrady-Steed, J., P.M. Harris and P.J. Morin (1997). Biodiversity regulates ecosystem predictability. *Nature* 390: 162-165.

Morgan, M.G. and M. Henrion (1990). *Uncertainty. A Guide to Dealing with Uncertainty in Quantitative Risk and Policy Analysis.* Cambridge, Cambridge Univ. Press.

Morton, R.D. and R. Law (1997). Regional species pools and the assembly of local ecological communities. *J. theor. Biol.* 187: 321-331.

Müller, F. (1997). State-of-the-art in ecosystem theory. *Ecological Modelling* 100(1-3): 135-162.

Müller-Wenk, R. (1997). *Safeguard Subjects and Damage Functions as Core Elements of Life-Cycle Impact Assessment*. St. Gallen, IWÖ-Diskussionsbeitrag 42. Institute for Economy and the Environment (IWÖ), University St. Gallen.

Müller-Wenk, R. (1998). *Land Use – The Main Threat to Species. How to Include Land Use in LCA*. St. Gallen, IWÖ-Diskussionsbeitrag 64, available at http://www.iwoe.unisg.ch/service/veroeff/db64.htm. Institute for Economy and the Environment (IWÖ), University St. Gallen.

Naeem, S., K. Håkansson, J.H. Lawton, M.J. Crawley and L.J. Thompson (1996). Biodiversity and plant productivity in a model assemblage of plant species. *OIKOS* 76: 259-264.

Naeem, S. and S. Li (1997). Biodiversity enhances ecosystem reliability. *Nature* 390: 507-509.

Norgaard, R.B. (1989). The case of methodological pluralism. *Ecological Economics* 1: 37-57.

O'Neill, R.V. (1989). Perspectives in hierarchy and scale. In: *Perspectives in Ecological Theory*. Roughgarden, J. and R.M. May, Princeton, New Jersey, Princeton University Press: 140-156.

OECD (1996). *Saving Biological Diversity. Economic Incentives*. Paris, Organisation for Economic Co-operation and Development (OECD).

OECD (1998). *Towards Sustainable Development. Environmental Indicators*. Paris, Organisation for Economic Co-operation and Development (OECD).

Palisade Corporation (1997). *Benutzerhandbuch für @RISK. Risikoanalysen- und Simulations-Add-In für Microsoft Excel und Lotus 1-2-3*. Newfield, Palisade.

Pearce, D. and D. Moran (1994). *The economic value of biodiversity*. London, Earthscan.

Perrings, C., et al., Ed. (1995). *Biodiversity loss: Economic and ecological issues*. Cambridge, New York, Melbourne, Cambridge Univ Press.

Perrings, C.A., K.G. Mäler, C. Folke, C.S. Holling and B.-O. Jansson, Eds. (1995). *Biodiversity Conservation: Problems and Policies*. Dordrecht, Kluwer.

Pfiffner, L., J.-M. Besson and U. Niggli (1995). DOK-Versuch: vergleichende Langzeituntersuchungen in den drei unterschiedlichen Anbausystemen biologisch-dynamisch, organisch-biologisch und konventionell. III. Boden: Untersuchungen über die epigäischen Nutzarthropoden, insbesondere über die Laufkäfer (Col. Carabidae), in Winterweizenparzelle. *Schweiz. Landw. Fo., Speciel edition DOK* 1: 1-34.

Preston, F.W. (1962). The canonical distribution of commoness and rarity: Part I and II. *Ecology* 43: 185-215, 410-432.

Reidl, K. (1989). *Floristische und vegetationskundliche Untersuchungen als Grundlage für den Arten- und Biotopschutz in der Stadt – dargestellt am Beispiel Essen.* PhD thesis, Essen, GHS Essen.

Rubik, F. and V. Teichert (1997). *Ökologische Produktpolitik. Von der Beseitigung von Stoffen und Materialien zur Rückgewinnung in Kreisläufen.* Stuttgart, Schäffer-Poeschel.

Schaefer, M. (1992). *Wörterbuch der Biologie. Ökologie.* Jena, Gustav Fischer.

Schaefer, M. (1996). *Die Bodenfauna von Wäldern: Biodiversität in einem ökologischen System.* Mainz, Abhandlungen der Math-naturw. Klasse, Nr. 2. Akademien der Wissenschaften und der Literatur.

Schaltegger, S. (1997). Economics of Life Cycle Assessment: Inefficiency of the present approach. *Business Strategy and the Environment* 6: 1-8.

Schläpfer, F. and B. Schmid (1999). Expert estimates about effects of biodiversity on ecosystems processes and services. *OIKOS* 84(2): 346-352.

Scholz, R. and O. Tietje (in press). *Embeded Case Study Methods. Integrating Quantitative and Qualitative Knowledge.* Newbury Park, London, New Delhi, SAGE.

Schreiber, C. (1995). *Vergleich der Artenvielfalt von konventionellen-, IP- und Biobetrieben (auf verschiedenen Unterlagen in der kollinen-submontanen Stufe) im westlichen Aargauer Mittelland.* Diploma thesis, Zürich, Geobotanisches Institut und Institut für Agrarwirtschaft, ETH Zürich.

Schulte, W. (1985). *Florenanalyse und Raumbewertung im Bochumer Stadtbereich.* Bochum, Materialien zur Raumordnung, Band XXX. Geographisches Institut der Ruhr-Universität Bochum.

Schulze, E.D. and H.A. Mooney, Eds. (1994a). *Biodiversity and ecosystem function.* Berlin, New York, Springer.

Schulze, E.D. and H.A. Mooney (1994b). Ecosystem function of biodiversity. A summery. In: *Biodiversity and ecosystem function.* Schulze, E.D. and H.A. Mooney, Berlin, New York, Springer: 525.

Schumacher, W. (1980). Schutz und Erhalt gefährdeter Ackerwildkräuter durch Integration von landwirtschaftlicher Nutzung und Naturschutz. *Natur und Landschaft. Zeitschrift für Umweltschutz und Landespflege.* 55(12): 447-453.

Schweinle, J. (1998). Integration of land use into Life Cycle Assessment. In: *Proceedings of the International Conference on Life Cycle Assessment in Agriculture, Agro-Industry and Forestry.* VITO, Brussels.

Seidl, I. and J. Gowdy (1999). Monetäre Bewertung von Biodiversität: Grundannahmen, Schritte, Probleme und Folgerungen. *GAIA* 8(2): 102-112.

Sobol, I.M. (1994). *A Primer for the Monte Carlo Method.* Boca Raton, CRC Press.

Sprigge, T.L.S. (1997). Gibt es in der Natur intrinsische Werte? In: *Ökophilosophie.* Birnabcher, D., Stuttgart, Recalm.

Steen, B. (1997). *The EPS-System 1997 – a Comprehensive Presentation.* Göteborg. Chalmers University of Technology (CPM).

Steen, B. (1999). *A systematic approach to environmentla priority strategies in product development (EPS). Version 2000 – Models and data of the default method.* Göteborg, report no.1999:5. Chalmers University of Technology (CPM).

Steen, B. and S.O. Ryding (1992). *The EPS enviro-accounting method: an application of environmental accounting principles for evaluation and valuation of environmental impact in product design.* Göteborg, IVL-rapport. IVL.

Sukopp, H., Ed. (1990). *Stadtökologie. Das Beispiel Berlin.* Berlin, Dietrich Reimer.

Suter II, G.W., L.W. Barnthouse, S.M. Bartell, T. Mill, D. Mackay and S. Paterson (1993). *Ecological risk assessment.* Chelsea, Michigan, Lewis Publishers.

Swan, G., Ed. (1998). *Evaluation of Land Use in Life Cylce Assessment.* Göteborg, Centre for Environmental Assessment of Product and Material Systems. Chalmers University. Technology Report1998:2.

Tansley, A.G. (1935). The use and abuse of vegetational concepts and terms. *Ecology* 16: 284-307.

Tilman, D. (1997). Biodiversity and ecosystem functioning. In: *Nature's Services. Societal Depencendence on Natural Ecosystems.* Daily, G.C., Washington, D.C., Covelo, Island: 93-112.

Tilman, D., J. Knops, D. Wedin, P. Reich, M. Ritchie and E. Siemann (1997). The influence of functional diversity and composition on ecosystem processes. *Science* 177: 1300-1302.

Turner II, B.L. and W.B. Meyer (1994). Global land-use and land-cover change: An overview. In: *Changes in Land Use and Land Cover: A Global Perspective.* Meyer, W.B. and B.L. Turner II, Cambridge, Cambridge University Press.

Udo de Haes, H., Ed. (1996). *Towards a Methodology for Life Cycle Impact Assessment.* Brüssel, Society of Environmental Chemistry and Toxicology (SETAC).

Udo de Haes, H., O. Jolliet, G. Finnveden, M. Hausschild, W. Krewitt and R. Müller-Wenk (1999). Best available practice regarding impact categories and category inidcators in Life Cycle Impact Assessment. Background document for the second working group on Life Cycle Impact Assessment of SETAC-Europe (WIA-2) Part A. *Int. J. LCA* 4(2): 66-74.

UNCED (1992a). *Report of the United Nations Conference on Environment and Development.* Rio de Janeiro, 3-14 June 1992, United Nations Conference on Environment and Development (UNCED).

UNCED (1992b). Report of the United Nations Conference on Environment and Development. Section II: Conservation and Management of Resources for Development. Chapter 10: Integrated Approach to the Planning and Management of Land Resources. In: Rio de Janeiro, 3-14 June 1992, United Nations Conference on Environment and Development (UNCED).

UNEP (1992). *Convention on Biological Diversity*. Nairobi, Kenya, available at gopher://gopher.undp.org:70/00/unconfs/UNCED/English/biodiv; Date 4.02.99. United Nations Environment Programme (UNEP).

Welten, M. and R. Sutter (1982). *Verbreitungsatlas der Farn- und Blütenpflanzen der Schweiz*. Basel, Birkhäuser.

Wenzel, H., M. Hausschild and L. Alting (1997). *Environmental Assessments of Products. Volume 1: Methodology, tools and case studies in product development*. London, Chapman & Hall.

Wiertz van Dijk, J. and J.B. Latour (1992). *MOVE: Vegetatie-module; de kans op voorlomen van 700 plantsoorten als functie van vocht, pH, nutienten en zout*. Bilthoven, report no. 711901006. RIVM.

Wisheu, I.C. and P. Keddy (1996). Three competing models for predicting the size of species pools: a test using eastern North American wetlands. *Oikos* 76: 253-258.

Wittwer, A., R. Meier, P. Bolliger, J. Wittwer, P. Thomet, E. Thomet and H. Beyeler (1997). *Ökologischer Ausgleich. Erste Erfolgskontrollen in drei Regionen aus der Sicht der Förderung der Artenvielfalt*. Bern, Umwelt-Materialien Nr. 82, Natur und Landschaft. Bundesamt für Umwelt, Wald und Landschaft (BUWAL).

Wohlgemuth, T. (1992). Die vegetationskundliche Datenbank. *Schweiz. Z. Forstwes*. 143: 22-36.

Wohlgemuth, T. (1993). Der Verbreitungsatlas der Farn- und Blütenpflanzen der Schweiz (Welten und Sutter 1982) auf EDV: Die Artenzahl und ihre Abhängigkeit von verschiedenen Faktoren. *Bot. Helv*. 103: 55-71.

Wohlgemuth, T. (1998). Modelling floristic species richness on a regional scale: a case study in Switzerland. *Biodiversity and Conservation* 7: 159-177.

Wood, P.M. (1997). Biodiversity as the source of biological resources: A new look at biodiversity values. *Environmental Values* 6: 251-268.

WSL and BUWAL (1999). *Schweizerisches Landesforstinventar. Ergebnisse der Zweitaufnahme 1993-1995*. Birmensdorf, Bern, Eidgenössische Forschungsanstalt für Wald, Schnee und Landschaft (WSL), Bundesamt für Umwelt, Wald und Landschaft (BUWAL).

WSL/FNP (without year). *EDV-Flora der Schweiz 1.0*. Birmensdorf, Eidgenössische Forschungsanstalt für Wald, Schnee und Landschaft (WSL/FNP).

Zobel, M. (1997). The relative role of species pools in determining plant species richness: an alternative explanation of species coexistence? *Trends Ecol. Evol*. 12(7): 266-269.

Index